# THE METRIC SYSTEM

# THE METRIC SYSTEM

## A Laboratory Approach for Teachers

### Nancy C. Whitman
University of Hawaii

### Frederick G. Braun
University of Hawaii

John Wiley & Sons
New York   Santa Barbara   London   Sydney   Toronto

Copyright © 1978, by John Wiley & Sons, Inc.

All rights reserved. Published simultaneously in Canada.

No part of this book may be reproduced by any means, nor
transmitted, nor translated into a machine language without the written permission of the publisher.

*Library of Congress Cataloging in Publication Data:*

Whitman, Nancy C
    The metric system.

    Includes index.
    1. Metric system—Study and teaching (Elementary)—United States.
2. Physical measurements—Study and teaching (Elementary)—United States.
3. Physics—Experiments.   I. Braun, Frederick G., 1926–   joint author.
II. Title.

RC97.W49           530'.8          77-22793
ISBN 0-471-02763-4

Printed in the United States of America

10 9 8 7 6 5 4 3 2 1

# PREFACE

The purpose of this worktext is to educate the teacher to know, to do, and to think metric. This is done by providing teachers with activities in which they directly measure physical things using metric units. These activities do not provide for conversions from the customary units to the metric units and vice versa. Instead they encourage teachers to express equivalent metric measurements using different metric units. The activities encourage teachers to estimate in metric units. Many of the activities are readily adaptable to the teachers' classroom use. The numerous text exercises allow teachers to test their understanding of metrics.

The worktext is written for practicing elementary school teachers and for those future teachers who are not familiar with the metric system. The worktext can be used as a supplementary textbook in college mathematics courses for elementary school teachers or in elementary mathematics methods courses. For practicing elementary school teachers, the worktext serves as the basic textbook for workshops. The entire worktext can be studied in a 16-hour workshop. In 8-hour workshops, that emphasize the metric content, Chapters 1 to 7 will provide the instruction and practice needed.

An efficient way to conduct the workshops or class sessions is to set up a series (about five) of work and study stations in advance at which teachers explore and experiment with the materials. Task cards at the stations could specify the tasks and activities that should be carried out.

Most of the materials included in this book have been used for three years in metric workshops for elementary school teachers. A preliminary version of the book was tested out with preservice teachers. Both groups have found that the material enhances their understanding of metrication.

For preservice and inservice teachers who wish to extend their knowledge of the metric movement and for those who wish to develop lesson plans for classroom use, the bibliography in Appendix II provides a wealth of possibilities.

The novice who needs help in using the laboratory approach to teaching should find Chapter 9 "Setting Up a Classroom Laboratory for Teaching Metric Units" of value. Teachers who want to see how the metric units fit into an elementary mathematics program will find Chapter 8 "A Continuum of Metric Ideas" of interest. The teacher who is ready to purchase commercial supplies and equipment and other resource materials will find Appendix II of great assistance. Finally, the activities described in Appendix III can be reproduced in a classroom.

We wish to thank L. E. Barbrow of the National Bureau of Standards, Mildred Higashi of the Hawaii State Department of Education, and Annette N. Nishimura for their helpful criticisms and suggestions, Fay Zenigami for her assistance in the bibliographic review, and Robert Jurich for his valuable assistance in the typing.

Nancy C. Whitman
Frederick G. Braun

# CONTENTS

# EQUIPMENT AND SUPPLIES

The materials needed for performing the tasks and activities in the book are listed here. In some cases readers will be able to substitute similar items for those listed here. The materials lists are repeated in the chapters in which they are needed.

chapter three   Meter stick without graduations
                Adding machine tape
                Banana
                Coffee jar
                Soda pop bottle or can
                Roll of ribbon
                Meter stick
                Ball of twine
                Scissors
                Dictionary

chapter four    Yarn (a small ball)
                Large milk carton
                Cardboard or stiff paper (Oaktag)
                File card
                Old comics and newspaper
                Glue
                Postage stamp (new or used)
                Florist wire
                Thread
                Cuisenaire rods
                Construction paper (30 cm by 45 cm)
                Masking tape (about 1 cm wide)
                4 to 10 different size cartons (e.g., tissue, ice
                    cream, matches)
                Cubic centimeters

chapter five    Beakers: 1000 mℓ, 500 mℓ, 250 mℓ,
                     50 mℓ, 10 mℓ
                Large jar: Coffee, mayonnaise, pickle, or jam
                2 large, 2 medium, 1 small plastic or paper cup
                6 miscellaneous common containers (e.g., Coke,
                     pickle, medicine bottles)

chapter six    1 wire clothes hanger
                Butter or whipped cream plastic tubs
                Play-Doh or clay
                Elbow macaroni (2 cupfuls) or other uniform set
                     of objects of the same size
                Bottle caps (about 25)
                Paper clips
                Flashlight battery, size D
                Wooden clothespins
                Grain of raw rice
                Grain of salt
                1 two-pan balance and a set of weights
                Additional weights:
                    10 1 gram       10 hectogram
                    10 dekagram     1 kilogram

chapter seven    Celsius thermometers
                Ice
                Salt
                Hot and cold water
                A plastic cubic decimeter
                Individual cereal box
                Small coffee can

To Jo,
Briggs,
and Dana

# chapter
## one

# INTRODUCTION

*The Metric System* is a laboratory approach to learning about a system of measurement that has been officially adopted by most of the world's nations. By this approach to learning teachers are expected to have experiences in directly measuring physical things using the metric units. In addition to this practical experience teachers are led through stages of concrete applications and symbolic problem solving. Teachers may go through the stages of learning individually, in pairs, or in groups. As a check on their learning, most of the answers to symbolic problems are provided in the book.

To assist the teachers "to think metric" and "to feel metric," many experiences in "eyeballing" measurements are provided. In addition, the activities described encourage teachers to express equivalent metric measurements using different metric units. No experiences are provided in converting from customary units to metric units and vice versa. It is believed that being able "to think metric" and "to think customary" will provide the necessary knowledge for most conversions. For exact conversions between the two systems, it is much easier to use a mechanical converter which is now readily available.

The metric units covered in this book are length, capacity, mass, and temperature. The units are the ones studied at the elementary school level. In addition to the metric units themselves a brief review of the decimal system of numeration and of errors of measurement is provided. Before the metric units are introduced, the metric movement is reviewed and measurement in general is discussed.

To assist the classroom teacher in teaching the metric units, teaching strategy, classroom management, and student evaluation are discussed. Also provided are possible schemes for changing from a passive classroom to one that includes activities. To acquaint the classroom teacher with the place of metric concepts in a mathematics program, a continuum of these concepts is suggested.

The classroom teacher now beginning to teach the metric system, can refer to Appendix II for a listing of commercial supplies and equipment along with approximate prices and names and addresses of vendors. For teachers who wish to extend their backgrounds and enrich their classes, an annotated bibliography, and selected additional resources including task cards, audiovisuals, workbooks, and games are provided. Also provided are sample classroom activities.

Upon completion of this book, teachers should be able to:

1. State why it is the policy of the United States to plan the increasing use of the metric system.

2. Use the metric units of length, mass, capacity, and temperature.

3. Estimate using metric units.

4. Convert internally within the metric system.

5. Recognize the relationship of the decimal system of numeration to the metric system of measurement.

6. Find an appropriate place for metric concepts in a mathematics program.

7. Obtain and develop resource materials for metric education.

8. Use methodologies appropriate to teaching metric concepts and skills.

## GOING METRIC: A BRIEF HISTORY

### In France

In 1790, an edict of King Louis XVI ordered scientific investigation into a reform of French weights and measures. The result of this investigation — the metric system — had intellectual foundations in the rebirth of scientific interest in France between the sixteenth and eighteenth centuries. In 1670, Gabriel Mouton had proposed a comprehensive decimal system of weights and measures. He defined its basic unit of length as a fraction of the length of a great circle of the earth (i.e., a full line of latitude or longitude).

In 1795, France officially adopted the metric system. Copies of the provisional standards were sent to several countries including the United States. Use of the metric system was compulsory in 1840.

### In the United States

Secretary of State John Quincy Adams, in 1821, submitted an exhaustive report on the subject of weights and measures to Congress. Adams recommended retention of the English customary system by the United States. Several decades later (1866), Congress made the use of the metric system legal. It also provided that each state be furnished with a set of standard weights and measures of the metric system.

**Treaty of the Meter**

The Treaty of the Meter was signed by 17 nations, including the United States, in 1875. The treaty provided for the fabrication of new and improved standards for metric weights and measures, the establishment and maintenance of a permanent International Bureau of Weights and Measures, and the creation of a General Conference as a permanent deliberative body to pass upon matters affecting international weights and measures.

By 1880, 17 nations, including most of South America and the major European nations, had officially adopted the metric system, at least for government purposes. By 1900, 35 nations had accepted the metric system. In 1893 the United States prototype meter and kilogram became the nation's fundamental standards of length and mass. The units of English customary system were defined not by their own standards but by carefully specifying what fraction of a meter would constitute a yard and what fraction of a kilogram would constitute a pound.

**The International System of Units**

In 1960, at the General Conference on Weights and Measures, a new international standard of length, based on the wavelength of the element krypton was adopted in place of the original "metal bar." At the same time, the modernized system was officially renamed the Système International d'Unités (the International System of Units) or "SI."

**Great Britain Goes Metric**

In 1965, the president of the British Board of Trade, with the approval of the government, announced plans for the conversion of Great Britain to the metric system "sector by sector" over a 10-year period. The object of the changeover was to mesh British standards with those of Continental Europe, the largest market for British export. This move by Great Britain left the United States as the only major power using nonmetric units. This fact influenced events in the United States in the following decade.

### The "Metric Conversion Act of 1975"

In 1968, Public Law 90-472 was enacted in the United States. This law authorized the Secretary of Commerce

> to conduct a program of investigation, research, and survey to determine the impact of increasing worldwide use of the metric system on the United States; to appraise the desirability and practicability of increasing the use of metric weights and measures in the United States; to study the feasibility of retaining and promoting by international use of dimensional and other engineering standards based on the customary measurement units of the United States; and to evaluate the costs and benefits of alternative courses of action which may be feasible for the United States.

In 1971, the Secretary of Commerce transmitted to Congress the Report on the U. S. Metric Study. The report recommended "that the United States change to the International Metric System deliberately and carefully; that early priority be given to educating every American school child and the public at large to think in metric terms"; and "that Congress, after deciding on a plan for the nation, establish a target date ten years ahead, by which time the United States will have become predominantly, though not exclusively, metric."

Meanwhile in 1970 Australia launched its 10-year metrication program, New Zealand started its 7-year program, and Canada announced that metrication was a definite objective of Canadian policy.

Attempts by this country to enact metric legislation in 1972 and 1974 failed. However, numerous state legislatures and state school boards took formal action toward going metric. An interstate consortium on Metric Education, with members from 28 states and territories, met in 1974 to plan how the nation's educational institutions can best prepare Americans to understand and to use metrics.

In 1975, two million dollars was appropriated to the Office of Education of the Department of Health, Education and Welfare to establish a metric education program to support models and demonstration projects that would improve and expand metric education throughout the country. By the end of 1975, metric education programs were underway in all 50 states.

On December 23, 1975, the "Metric Conversion Act of 1975" became Public Law 94-168 in the United States. Its purpose was "to declare a national policy of coordinating the increasing use of the metric system in the United States," and "to establish a United States Metric Board to coordinate the voluntary conversion to the metric system." At long last, the United States is going metric!

## METRICS AND THE DECIMAL SYSTEM OF NUMERATION

One of the arguments in favor of the metric system is that it is a decimal system of measurement, the system on which our currency is based and on which our system of numeration is based. Because the system is decimal in nature, it facilitates calculations. Generally one needs only to add zeros or decide where to place the decimal point when computing with metric units. Before you finish studying this book, you will have an opportunity to experience this ease of computation.

In education, the study of the metric units will help elucidate and reinforce the elementary school pupil's understanding of our system of numeration. Let us briefly review this system.

Our system is called a *decimal* system because it uses ten digits: 0, 1, 2, 3, 4, 5, 6, 7, 8, 9; and each digit has an apparent value and a place value. For example, in the numeral "23," 2 has an apparent value of two and a place value of ten. Therefore "2" in "23" represents 2 X 10 or 20; that is, it is the apparent value times the place value. Each digit in a numerical expression has a place value. More specifically, each place value has ten times the place value to its right and one-tenth the place value to its left. Here are the place values from one million to one-millionth:

9  8  7  6  0  2  1  .  2  3  4  5  6  7

millions | one hundred thousands | ten thousands | thousands | hundreds | tens | ones | tenths | hundredths | thousandths | ten-thousandths | hundred-thousandths | millionths

Since each place value has ten times the place value to its right, whenever the decimal point is moved to the right one space, the place value of each digit is increased ten times.

For example:  2  5  .  0    2  5  0

tens | ones | tenths | hundreds | tens | ones

However, the place value of a digit can be increased ten times simply by being multiplied by 10.

 For example:  2.5 X 10 = 25

Hence, we have the rule:

When multiplying by powers of ten, the power (number of zeros after the "1" in the multiplier) indicates the number of places to move the decimal point to the *right*.

More examples:   2.05 X 10  = 20.5

32.5  X 100 = 3250

0.65 X 100 = 65

Each place has one-tenth the place value to its left.  Hence, when the decimal point is moved to the left one space, the place value of each digit is decreased tenfold.

Examples:  3  .  6      0  .  3  6

0  .  1  7  8      0  .  0  1  7  8

However, the place value of the digit can be decreased tenfold only by being divided by 10.

For example:  2.5 ÷ 10 = 0.25
0.29 ÷ 10 = 0.029

Therefore, we have the rule:

Whenever dividing by powers of ten, the power (number of zeros after the "1" in the divisor) indicates the number of places to move the decimal point to the *left*.

More examples:  257 ÷ 10 = 25.7
28.9 ÷ 100 = 0.289
0.73 ÷ 100 = 0.0073

### Just for Practice

Fill in the blanks.

1. 235 X 10 = _____

2. 235 ÷ 10 = _____

3. 235 ÷ _____ = 2.35

4. 2.35 X 100 = _____

5. 2.35 ÷ 10 = _____

6. 2.35 ÷ _____ = 0.0235

7. 3.6 ÷ 100 = _____

8. 3.9 ÷ 1000 = _____

9. 3.7 X _____ = 370

10. 0.70 X 1000 = _____

11. 0.70 ÷ _____ = 0.0070

12. 0.09 ÷ 10 = _____

13. 0.010 ÷ 100 = _____

14. 0.002 ÷ _____ = 0.0002

15. 0.0020 ÷ 1000 = _____     16. 0.002 X 10 = _____

17. 0.002 X _____ = 2     18. 0.0263 ÷ 100 = _____

19. 0.073 ÷ _____ = 0.073     20. 0.0230 ÷ 1000 = _____

Here are the answers

| | |
|---|---|
| 1. 2350 | 2. 23.5 |
| 3. 100 | 4. 235 |
| 5. 0.235 | 6. 100 |
| 7. 0.036 | 8. 0.0039 |
| 9. 100 | 10. 700 |
| 11. 100 | 12. 0.009 |
| 13. 0.000 10 | 14. 10 |
| 15. 0.000 002 0 | 16. 0.02 |
| 17. 1000 | 18. 0.000 263 |
| 19. 1 | 20. 0.000 023 0 |

## EXERCISES

1. Obtain a copy of the "Metric Conversion Act of 1975" and report on (a) why the United States is committed to a policy of increasing use of the metric system, and (b) the act's provisions.

2. Read and report on *A Metric America: A Decision Whose Time Has Come* by Daniel V. DeSimone.

3. List the advantages for the United States in going metric.

4. Describe the metric conversion activities in Canada, Great Britain, and New Zealand. What role did the schools play in each country?

5. Make a list of things and circumstances where metric units are used.

6. In what country was the metric system originally introduced?

7. What does "SI" mean?

8. What were the provisions of the Treaty of the Meter?

9. Research and report on your school district's plans for metric education.

10. Why do you think conversion to metric is being delayed?

11. What are some of the problems involved in conversion to metric?

12. What special interest groups do you think object to conversion to metric? Why might they object?

# chapter
# two

# AN INTRODUCTION
# TO MEASUREMENT

In general, measurement is a process of assigning a number to some property of an object. Figure 2.1 provides examples of this process applied. In example *a* the property of quantity is assigned the number 5, in example *b* the property of length is assigned the number 4, and in example *c* the property of capacity is assigned the number 3. In each example, the assignment of the number could result from counting. However, what is chosen to be counted differs in each example. In example *a* the number of stars is counted, in example *b* the number of paper clips is counted, and in example *c* the number of cupfuls is counted. What is chosen to be counted is referred to as a unit of measurement. When the unit of measurement is agreed upon by a large segment of the people, it is called *a standard unit of measurement.*

Example *a* differs from examples *b* and *c* in that the objects counted in example *a* are separable, whereas those in examples *b* and *c* are not. Measurements as in example *a* are *discrete measurements* and those of examples *b* and *c* are *continuous.*

Continuous measurements are always only approximate. For once

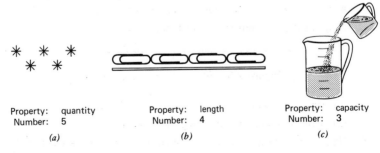

Property: quantity
Number:  5

Property:  length
Number:  4

Property:  capacity
Number:  3

(a)                    (b)                    (c)

**Figure 2.1**  Examples of the measuring process.

a unit of measurement is chosen, it is always theoretically possible to select a second unit of measurement smaller than the original chosen. Hence it is not possible to obtain an exact measurement, since a "more exact" measurement could be obtained by choosing a smaller unit of measurement. To further clarify, suppose you measured the segment shown in Figure 2.2 as precisely as you can using 1/8 inch as the unit of measurement. Now I can take one-half of the unit of measurement you used, that is, 1/16 inch, and use it as my unit of measurement. This will enable me to make a more precise measurement than yours. Clearly, the smaller the unit of measurement you chose, the more precise your measurements will be.

**Figure 2.2**   Line segment.

Not all objects are measured directly. That is, the unit of measurement is not placed directly on the object being measured. Instead it is placed on some other object which provides the desired measurement. A reading from a thermometer is an example of an indirect measurement. Temperature is not measured directly, instead the length of a column of liquid is measured. However, the length of the column of liquid is an accurate indication of the temperature. Measurements are either direct or indirect. At times direct measurements are not feasible or possible, and indirect measurements are often more convenient.

*Exercises*

1. Explain why counting is a form of measurement.
2. Describe continuous measurement in your own words. Give examples of some continuous measurements.
3. Measure the segment shown in Figure 2.3.

**Figure 2.3**   Segment to be measured.

   (a)  What unit did you use?
   (b)  What number did you assign the segment?
4. Name some standard units of measurement you have used in the past.

5.  Give an example and explain why area measurements are approximate.
6.  Order the following units of measurement from least precise to most precise:  your foot, your palm, your thumb.
7.  Name a part of your body that would be a more precise unit of measurement than your thumb.
8.  Give some examples of indirect measurements.

## THE LEARNING OF MEASUREMENT

If by counting a student correctly assigns the number five to each of the sets of beads shown in Figure 2.4, one would be inclined to say

Set *A*                                        Set *B*

**Figure 2.4**   Sets of beads.

the student can, and understands how, to measure discrete objects. However, if the student insists that Set *B* contains more beads than Set *A*, one would hesitate to make such an assertion.  The studies of Piaget have shown that at certain stages of cognitive development students have not grasped the notions of invariance or conservation of quantity.  For such students, quantity is the space occupied by the configuration.  What they see triumphs over the idea of invariance of quantity.  The idea that numerical quantity by its nature is invariant is not understood by the student.  Hence any attempt "to teach" measurement of discrete objects is fruitless.  Students need instead experiences that cause them to realize this property of quantity. Only then will it make sense to teach students how to measure discrete objects.  At age seven about 75 percent of the students come to realize the invariance property of quantity.

The notion of conservation is also fundamental to the development of continuous measurement concepts.  To perform linear measurements students must have developed the concept of conservation of length.  That is, when given two objects equal in length they are able to state that both still have the same length when one of them is

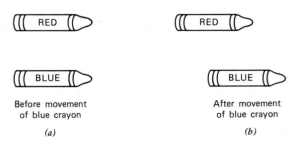

Before movement
of blue crayon

After movement
of blue crayon

*(a)*

*(b)*

**Figure 2.5**   Testing for conservation of length.

moved.  For example, in Figure 2.5, if students respond that the blue
crayon after is has been moved is longer than the red crayon, then
they have not developed the notion of conservation of length; if on
the other hand, they respond that the red and the blue crayons are
of equal length, then they have developed the notion of conservation
of length.  *If students have not reached this developmental stage, they
cannot learn to measure linearly.  Measuring length demands that the
student mentally construct an independent reference system in space
in which objects can be moved to new positions, but in which the
length of the object in each position in space does not change.*  Ac-
cording to Piaget, students understand conservation of length some-
time between ages six and eight.  Students who do not understand
this concept should be given experimental or trial-and-error experiences that involve making direct comparisons between objects.  Students would be asked to make simple matchings and comparisons and
to order objects.  Examples of these experiences follow:

*Example 1:*
Have a student collect five objects and order them by size from smallest to largest.

*Example 2:*
Give a student three pieces of tape of different length.  Have the student do the following:

   1.  Name five objects in the classroom *longer* than one of the
pieces of tape.

   2.  Name five objects in the classroom *shorter* than a second piece
of tape.

   3.  Name five objects in the classroom the *same* length as the last
piece of tape.

A student's ability to measure area is also related to his or her ability to conserve. To measure regions, students must have developed the concept of conservation of area. That is, when given two regions of equal area and shape, students are able to state that both of them have the same area, even if one of them is shaped differently. For example, in Figure 2.6 the student should state that regions *a* and *b* have the same area before and after region *b* is transformed by cutting and rearranging.

Region *a*

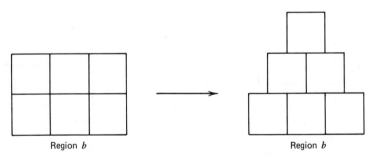

Region *b*          Region *b*

**Figure 2.6**  Conservation of area.

The student must realize that area does not necessarily change with shape. As stated previously, measurement consists of assigning a number to some property of an object. Basically this assignment takes the form of counting. Hence, since conservation of area is the ability to compare two regions of different shapes by counting the number of basic subunits of each, using a basic unit of measure, conservation of area is a prerequisite to measuring regions.

The basic process of measuring regions is illustrated in Figure 2.7. The basic unit is shown at the right. The property to be measured is area and the number assigned is 8.

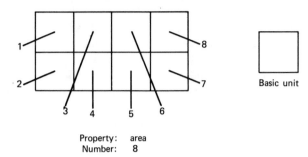

Property:   area
Number:    8

**Figure 2.7**   Basic process of measuring regions.

In Figure 2.8, if students count regions *x* and *y* each to be six units, and then assert that region *y* has a larger area, what they "see" has overwhelmed their intellectual idea of conservation of area. Generally speaking, students can conserve area about the time they conserve length. Furthermore, an understanding of conservation of length is not a prerequisite to the conservation of area. The conservation of area and length may develop simultaneously or either one may develop first.

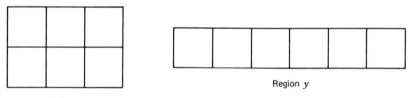

Region *x*

Region *y*

**Figure 2.8**   Testing for conservation of area.

For those students who are not able to conserve area, the teacher should provide readiness exercises. These experiences would involve making direct comparisons between objects. Following are some examples.

*Example 1:*
Have students form as many different interesting shapes as they can using all the tangram puzzle pieces (see Figure 2.9). Have them trace around the shapes they make.

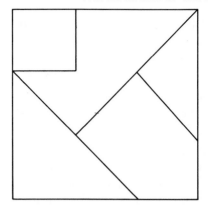

**Figure 2.9**  Tangram puzzle.

*Example 2:*

Give the student two pieces of grid paper of equal area and shape. Have the student cut out and arrange all the unit squares into whatever configuration he or she chooses. Then ask if the size or area of the configuration is the same as the region that was not cut out.

*Example 3:*

Give students a grid of square regions and different sized and shaped cards. Have students use their grid of square regions, which may be cut out, to determine which cards are

1. the same area as their grid of square regions
2. a larger area than their grid of square regions
3. a smaller area than their grid of square regions

*Exercises*

1. Give an example of what is meant by conservation of length.
2. With several six- or seven-year-old students, test the notion of conservation of quantity. Report your findings to the class.
3. Generally speaking, at about what grade level can a student conserve length?
4. Devise a test to determine if a student has developed the notion of conservation of length.
5. With several first- and second-grade students, try out a test designed to assess their notion of conservation of length.

6. Design an activity to determine if a student has developed the notion of conservation of area.
7. With elementary school age students try out an activity designed to determine their idea of conservation of area.
8. Design and try out a readiness activity for area measurement.
9. With children, try out the readiness activities described in Examples 1, 2, and 3 above.

To measure capacity students must develop the concept of conservation of liquid. That is, given two equal amounts of liquid, they can state that the amounts remain equal when one of them is poured into a different shaped container. That a student has developed this conservation concept can easily be tested by having students fill two identical containers with the same amount of water. Then have them pour water from one container into a third container which is much taller or wider (as shown in Figure 2.10). Ask them if the third container contains the same amount, less (or more) than the other container.

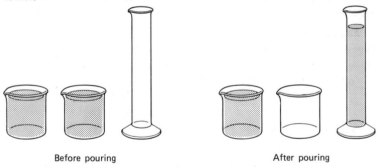

Before pouring                              After pouring

**Figure 2.10**  Testing for conservation of liquid.

Generally students will have attained this concept about the time they do the conservation of length and area. The teacher should provide readiness activities for those students who have not reached this level of development. Again these experiences would involve making simple matchings and comparisons. Some examples follow:

*Example 1*
Provide students with pairs of containers of different shapes and sizes. Have them decide which one will hold more. Have them verify by

filling it with water and then pouring it into the other container. Be sure to provide containers of different shapes that contain the same amount of liquid.

### Example 2

Bring in different sized juice cans. Have students arrange them from the least to the most liquid each can hold.

Understanding the idea of conservation of displaced or occupied volume is basic to measuring volume. That is, given two objects of the same shape and size but of different density, the student understands and can explain why both displace the same amount of water. Beginning at about age 12 or 13, students develop the ability to view volume as occupied space, understand the notion of a unit of volume, and see volume in terms of changes in linear measures of length, width, and height. This suggests that activities at the elementary school would focus on providing readiness experiences. For example, students might be asked to construct a tower on a 3 X 4 block base that has the same volume as another tower on a 3 X 2 block base. The development of the volume formula would wait until the junior high school years.

In order to measure mass (weight) students must have developed the idea that two equal masses remain equal when changed into different shapes, that is, the concept of conservation of mass is developed. The following activities and questions can help you determine whether students have reached this level of development.

### Activities

Have students make two balls of clay of equal mass. Have them use a two-pan balance to check. Ask them to roll one of the balls into the shape of a sausage. Now ask if the sausage has the same mass as the ball. Next have the students break the sausage into pieces and roll each into a ball. Ask if the combined mass of the small balls will have the mass of the big ball.

Generally, students will have developed the notion of conservation of mass about age 10 or 12. Before that age, activities should be confined to pre-mass measurement activities. These activities should be of direct comparison of objects. The following are some suggestions:

1. Students can compare mass by lifting equal volumes of unlike

materials, for example: Two empty cans of juice, one filled with sand and one with water.

2. Students can compare volumes of the same materials, for example: Two very different sized clay balls may be compared by lifting.

3. Students can make relative comparisons of two objects by placing both on a two-pan balance, for example: Have students determine which is heavier: (a) a pencil or a crayon, (b) a nickel or a quarter, (c) a styrofoam ball or clay ball.

4. Students can order the mass of objects. For example, Have students fill a set of nesting toys with water (plastic hexagons are good). Have them arrange the nesting containers in order starting with the one that holds the least amount of mass of water.

To understand the nature of temperature measurement students need premeasurement experiences with objects and situations of varying temperatures. They also need experiences designed to show how a thermometer works. Following are examples of simple experiments:

### Example 1
Fill two glasses with water: one warm and one cold. Have students determine which is colder by touch.

### Example 2
Have students make a list of things in the room that feel cool and a second list of things that feel warm.

### Example 3
Present a collection of pictures to the students and have them, using visual cues, arrange the pictures in order from hot to cold.

### Example 4
Have students make a simple thermometer as follows: Fill a bottle (one with a small opening) with water containing ink or food color. Insert a stopper containing a glass or plastic tube into the bottle so that some of the water shows in the tube (see Figure 2.11). Place the bottle into a container of boiling water.

About 15 minutes later, note and mark the water level. Now place the bottle into a container of ice. Again about 15 minutes later, note and mark the water level. Discuss "expansion" and "contraction" of liquids. Relate these to things in the students' environment, for example, ice melting on the lake in the springtime. Have stu-

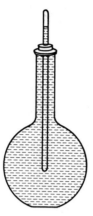

**Figure 2.11**   Simple thermometer.

dents refine their temperature scale by having them measure the temperature of other things using their simple thermometer, for example:
aquarium, refrigerator, drinking water.

*Exercises*

1. Give an example of what is meant by conservation of liquid.
2. Test some primary school children to assess if they have developed
   the notion of conservation of liquid.
3. Give an example of what is meant by conservation of mass.
4. Plan an activity to test the notion of mass conservation.
5. Plan a readiness activity for mass.
6. Plan a readiness activity for conservation of liquid.
7. Plan a readiness activity for conservation of volume.
8. Read and report on the history of the development of the Fahrenheit and Celsius thermometers.
9. Make a simple thermometer.
10. Research and report on the history of mass measures.

## OBJECTS, LANGUAGE, AND SYMBOLS OF MEASUREMENT

Students must practice measuring if they are to grasp the concept.
Learning experiences with measuring should take into consideration
the interrelationship of the objects of measurement, the verbalizations

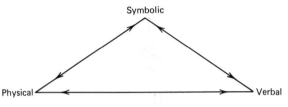

**Figure 2.12** Triangular model.

(oral and written), and symbols of measurement. The objects of measurements are the physical entities, for example, string and blocks. The verbalizations of measurement are written or oral phrases such as "ten centimeters," "twelve inches," "three feet," or "one-tenth of a mile." Mathematical symbols such as: 10, 12, 3, or 0.1 are completely lacking in verbalizations of measurement. The symbols of measurement are notations such as: 10 cm, 12", 3', and 0.1 mi. These notations can be expressed orally so that the focus is on the symbols. For example "10 cm" can be read as "one, zero, c, m" and "0.1 mi" may be read as "zero, point, one, mile."

Not infrequently in the past, books have emphasized the relationship of written verbalizations to their symbolization. Furthermore this was a one-way relationship in that written verbalizations were symbolized but symbols were not expressed in words. For example, students are asked to express "two thousand five and two-tenths" with numerical symbols, but they are not asked to express "2005.2" with written words.

The triangular model shown in Figure 2.12 focuses on the interrelationship of physical entities, verbalizations, and symbolizations of measurement. In effect, the model demonstrates the need for six types of activities. These are:

1. Physical measurements expressed in verbal terms.
2. Physical measurements expressed symbolically.
3. Symbolic measurements expressed verbally.
4. Symbolic measurements carried out physically.
5. Verbal expressions of measurement written symbolically.
6. Verbal expressions of measurement carried out physically.

There is great danger that rote learning will occur if any one of these activities is eliminated.

**Figure 2.13**  Segment to be measured.

Following are examples of each of the six types of activities.

1. From the *physical* to the *verbal.*

Have students measure lengths of objects and then express the measurement in words:  Measure Figure 2.13.  Express the measurement in words:  two inches.

2. From the *physical* to the *symbolic.*

Have students measure the length of objects and record their measurements symbolically.

A student measures a piece of tape that is two inches long and records:  2".

3. From the *symbolic* to the *verbal.*

Have students read out measurements expressed symbolically, for example:

| | |
|---|---|
| 2' | "two feet" |
| 0.2 mi | "two-tenths of a mile" |
| 3 cm | "three centimeters" |

4. From the *symbolic* to the *physical.*

Give students some symbolic measurements, for example: 10 cm, 11", 2 g, and ask them to measure various objects in the classroom to determine which have the values provided.

5. From the *verbal* to the *symbolic.*

Provide students with measurements expressed in words (in writing or orally).  Then have the students express their measurements in symbols.  For example:

Express these measurements in symbolic form:

| | Acceptable Answers |
|---|---|
| ten centimeters | 10 cm |
| twelve inches | 12" |
| one tenth of a mile | 0.1 mi |

6. From the *verbal* to the *physical.*

Have students measure or construct objects in the classroom with measurements stated verbally.

"Cut pieces of tape with the following measurements:  ten centimeters, twelve inches, one tenth of a meter."

*Exercises*
1. Review an elementary textbook series to determine which of the six types of activities suggested by the triangular model are provided.
2. Plan measurement activities that reflect the activities suggested by the triangular model.
3. Try out the activities planned in number 2 above with elementary school children.

## ESTIMATION

Estimation is a central idea of mathematics education. It is a process whereby one could determine the reasonableness of results of computing, measuring, or solving problems. Throughout this section focus will be on estimating measurements instead of on computing or solving problems.

Estimation is essentially a mental process. It enhances the learning of the measuring process and of the units of measurement because it focuses on the unit of measurement and the thing being measured; it reinforces the unit being used. Because estimating is essentially a logical process, teachers should consider the student's ability to be logical before requiring the student to make estimates. For example, a student who has not attained adult logic might "reason" that he or she is heavier than a classmate because the teacher said so.

The "tools" for estimating are "eyes" or their "extensions" (that is, body measures and relative sizes) when the unit of measurement and the object whose size is being estimated are present. Here estimating might be referred to as "eyeballing." However, when the object whose measurement is being estimated and the unit of measurement are both absent, all of the "tools" are mental images. Initially students should be asked to estimate when both the unit of measurement and the object whose measurement is being estimated are present. Then they should have experiences wherein either the unit of measurement or the object is present. And lastly, students should experience estimating when both the unit of measurement and the object are absent. Examples of these types of experiences follow.

1. Both the unit of measurement and the object whose measurement is being estimated are present.

Have students look at meter tapes or sticks. Then ask them to estimate to the closest meter the length of the classroom.

2. Only the unit of measurement is present.

Have students look at their meter sticks or tapes. Then ask them to estimate to the closest meter the distance from the classroom to the principal's office.

3. Only the object being estimated is present.

Without using their meter tapes or sticks, have students estimate to the closest decimeter the height of the teacher's desk.

4. Both the unit of measurement and the object being estimated are absent.

Without using their meter sticks or tapes, ask students to estimate the heights of their parents.

Generally when one thinks of estimating measurements, one thinks of a unit of measurement and the object being estimated instead of the reverse process, that is, given a measurement, name an object with those measurements. In this process, again the physical entities, that is, object(s) and unit of measurement, could both be present, only the object(s) or unit of measurement could be present, or neither the object(s) nor the unit of measurement could be present. Following are examples of classroom activities of these different situations.

1. Both the unit of measurement and the object concerned are present.

Have students look at their liter container, and then have them name several objects in the classroom that may contain two liters.

2. Only the unit of measurement is present.

Have students lift a kilogram unit, then have them name an object at home with the mass of one kilogram.

3. Only the objects being considered are present.

Given an orange, can of soup, and baseball, have students determine which object is closest in mass to one kilogram.

4. Both the unit of measurement and the object concerned are absent.

With capacity units absent, have students name objects outside of the classroom which have capacity of: 100 mℓ, 250 mℓ, and 500 mℓ.

## EXERCISES

1.  Characterize the estimating process as applied to measurement situations.

2.  Of what value is it to have students estimate measurements?

3.  Make up activities for primary school children wherein they experience the following kinds of estimation situations.

  a.  The measurement is given and the unit of measurement is present. The object(s) concerned is absent.

  b.  The measurement is given and the object(s) concerned is present. The unit of measurement is absent.

  c.  The measurement is given and both the unit of measurement and the object(s) concerned are present.

  d.  The measurement is given and both the unit of measurement and the object(s) concerned are absent.

  e.  Both the unit of measurement and the object being estimated are present.

  f.  The unit of measurement is present, and the object being estimated is absent.

  g.  The object whose measurement is being estimated is present, and the unit of measurement is absent.

  h.  Both the unit of measurement and the object being estimated are absent.

# chapter
# three

# LINEAR
# MEASUREMENT

*TASK 1*

To obtain experience in using the meter.

*Materials*

meter stick without graduations

As previously stated, the advantage of using the metric system is the ease with which computations can be made because of its decimal base. The *meter,* symbolized m, is the basic linear unit and each unit larger than a meter is a ten-fold multiple of the meter. A unit smaller than the meter is a decimal fraction of the meter.

Originally the meter was defined as one ten-millionth of the distance between the north pole and the equator. Later on when methods of measuring improved, it was shown that the original measurements to determine the meter length were in error, and hence the meter was redefined as the distance between two marks on a metal bar constructed from the original measurements. This metal bar became the meter prototype. Today the meter is defined as 1 650 763.73 wavelengths of the orange-red light emitted by a krypton 86 atom in a vacuum.

This chapter contains activities and exercises that will familiarize the participant with the relationships between various units and the meter. These activities should enable you to visualize and use the new linear units.

*Activity 1*

Relating meter length to self.

1. Measure your arm span. Place one end of a meter stick at the center of your chest and extend the stick out to your fingertips. How close to a meter is your arm span?

2. Measure your height. Are you two meters tall?

3. Lay the meter stick on the floor. About how many of your foot lengths make a meter?

4. While the meter stick is on the floor set your pace so that your stride is approximately one meter. Now pace off the length of the room. Next, pace the width. Now check pacing by measuring with a meter stick.

### Activity 2

Estimating and verifying the linear measure of common things to the closest meter.

1. Estimate and record in the table below the linear measure of the following:
   - a. Width of a hallway
   - b. Length of a seesaw
   - c. Distance you can throw a crumpled paper
   - d. Height of a jungle gym
   - e. Distance from here to the next room
   - f. Height of a water fountain

Table 3.1 Estimating and Measuring to the Closest Meter

| Object | Estimate | Measurement |
|---|---|---|
| a. Width of a hallway | | |
| b. Length of a seesaw | | |
| c. Distance you can throw a crumpled paper | | |
| d. Height of a jungle gym | | |
| e. Distance from here to the next room | | |
| f. Height of a water fountain | | |

2. By actually measuring, determine to the closest meter, the measurement of things listed under (1). Record your answers in the table above.

### Activity 3
Relating meter length to the outdoors.

1. With a partner, go outdoors and find three objects of interest to measure to the closest meter.

2. Estimate and then measure the three objects.

3. Name three objects outdoors that have an approximate length of one meter.

### Activity 4
Relating meter length to the home.

1. Name three objects in your home that are approximately one meter long.

2. Measure three objects in your home to the closest meter.

3. Bring these objects to class to be measured to the closest meter. Have someone in class measure these objects. Check the measurements.

## TASK 2
To obtain experience in using the decimeter.

### Materials
adding machine tape, meter stick without graduations

### Activity 1
Relating common objects to the decimeter.

1. Cut a piece of adding machine tape the length of the line in Figure 3.1.

---

**Figure 3.1** One decimeter.

a. This length is called a *decimeter,* symbolized dm.
b. How many of these lengths make a meter? Check your guess against a meter stick.
c. Do you see why "deci" means "0.1" and "decimeter" means "0.1 of a meter"?
d. Measure to the closest decimeter the length and width of this sheet

e. Measure to the closest decimeter the length of the shortest and longest sentence on this page.

### Activity 2
Estimating and verifying the linear measure of common objects to the closest decimeter.

1. Estimate, to the closest decimeter, and record in the table below the linear measure of each of the following objects:
   - a. Length of a banana
   - b. Height of a coffee jar
   - c. Width of a facial tissue
   - d. Width of the palm of your hand
   - e. Height of a soda pop can or bottle
   - f. Length of your shoes

**Table 3.2** Estimating and Measuring to the Closest Decimeter

| Object | Estimate | Measurement |
|---|---|---|
| a. Length of a banana | | |
| b. Height of a coffee jar | | |
| c. Width of a facial tissue | | |
| d. Width of the palm of your hand | | |
| e. Height of a soda pop can or bottle | | |
| f. Length of your shoes | | |

2. By measuring with your paper decimeter, determine to the closest decimeter, the measurement of the objects listed under (1). Record your measurements in the table above.

### Activity 3
Relating the decimeter length to the outdoors and to the home.

1. Name three things outdoors that have an approximate length of one decimeter. Check by measuring.

2. Name three things at home that are approximately one decimeter long. Check by measuring.

3. To the closest decimeter estimate and then measure the following.

    a. Length of a car

    b. Length of your bathtub or shower

    c. Height of your stove

    d. Width of your refrigerator

    e. Length of your mailbox

### Activity 4

Measuring and recording measurements equal to or greater than 10 decimeters but less than 100 decimeters.

Since 10 decimeters equal 1 meter, whenever measurements exceed or equal 10 decimeters, one might think to express the measurements in meters and decimeters. For example, if your height was 16 decimeters, you might think to express this as 1 meter and 6 decimeters. However, a more efficient way to express it is "one and six-tenths meters" which is symbolized "1.6 m". This latter method is recommended and commonly used.

1. Measure the following to the closest decimeter. Record your measurements in abbreviated form.

    a. Length of a roll of ribbon

    b. Length of a bookshelf

    c. Height of the slide

    d. Combined length of 10 books

    e. Combined length of 10 $1 bills ($5 and $10 will do also!)

2. Write these measurements in abbreviated form using meter as the unit of length. The first two are done for you.

    a. 99 decimeters = 9.9 m

    b. 6 meters and 7 decimeters = 6.7 m

    c. 87 decimeters

    d. 2 meters and 3 decimeters

    e. 18 dm

    f. 5 m and 3 dm

    g. 7 and eight-tenths meters

    h. 55 decimeters

    i. 78 dm

    j. 9 and three-tenths meters

Did you get these answers?

c.  8.7 m                    g.  7.8 m
d.  2.3 m                    h.  5.5 m
e.  1.8 m                    i.  7.8 m
f.  5.3 m                    j.  9.3 m

### TASK 3

To obtain experience in using the centimeter.

#### Materials

meter stick, adding machine tape, paper decimeter

A commonly used linear unit is the ***centimeter,*** symbolized cm, because many of our personal objects can best be measured in terms of centimeters. Although the centimeter cannot provide the precision needed in science and technology, it is a "handy" unit for most people estimating lengths of small objects. If you can identify some part of your hand that is approximately a centimeter in length, this will give you a handy reference until you have learned to "think metric."

#### Activity 1

Relating common objects to the centimeter.

1. Use the centimeter shown in Figure 3.2 to divide your paper decimeter into centimeters.

———

**Figure 3.2** One centimeter.

    a. How many of these centimeters make a decimeter?
    b. How many centimeters make a meter?  Check your answer against a meter stick.
    c. Do you see why "centi" means "0.01" and "centimeter" means "0.01 of a meter"?

2. Measure the following to the closest centimeter:
    a. Length of your pencil
    b. Length of your little finger
    c. Width of your thumbnail
    d. Width of your wallet or coin purse
    e. Width of your notebook or textbook

*Activity 2*

Estimating and verifying the linear measure of common objects to the closest centimeter.

1. Alternate with your neighbor drawing lines and see if you can guess their lengths in centimeters. Repeat until you can guess correctly four out of five times.

2. Try estimating the length of three different objects. Now check your guess by measuring.

3. Estimate the sum of the lengths of the sides (perimeter) of Figures 3.3 *a* to *d* below. Record estimates. Check your estimates by measuring to the closest centimeter.

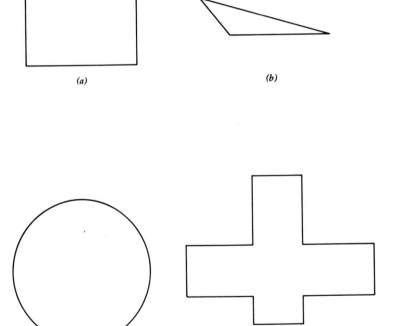

Figure 3.3 Estimating perimeter.

| Estimate | Measurement |
|---|---|
| (a) _____ cm | _____ cm |
| (b) _____ cm | _____ cm |
| (c) _____ cm | _____ cm |
| (d) _____ cm | _____ cm |

4. Estimate the combined lengths of a penny, nickel, dime, and quarter. Check your estimate by measuring.

### Activity 3
Relating the centimeter length to the outdoors and to the home.

1. Name three things outdoors that have an approximate length of one centimeter. Check by measuring.

2. Name three things at home that have an approximate length of one centimeter. Check by measuring.

3. To the closest centimeter measure the following.
   a. The diagonal width of a television screen
   b. The height of a table lamp
   c. The height of a fire hydrant
   d. The height of a plant
   e. The length and width of a bath towel

### Activity 4
Expressing centimeter measures in equivalent decimeter and meter form.

Since 10 centimeters = 1 decimeter, 1 centimeter = 0.1 decimeter. Also since 100 centimeters = 1 meter, 1 centimeter = 0.01 meter.
Hence we can say:

1 centimeter = 0.1 decimeter = 0.01 meter

or in symbolic form:

1 cm = 0.1 dm = 0.01 m

1. Let us use the above information to fill in the blanks.

   a.   7 cm = _____ dm = _____ m

   b.   12 cm = _____ dm = _____ m

   c.   185 cm = _____ dm = _____ m

   d.   105 cm = _____ dm = _____ m

   e.   999 cm = _____ dm = _____ m

   f.  1002 cm = _____ dm = _____ m

Did you recognize any patterns in your answers?  Did you get these
answers?

|       |         |        |
|-------|---------|--------|
| a.    | 0.7     | 0.07   |
| b.    | 1.2     | 0.12   |
| c.    | 18.5    | 1.85   |
| d.    | 10.5    | 1.05   |
| e.    | 99.9    | 9.99   |
| f.    | 100.2   | 10.02  |

  2.  If you need more practice try these:

     a.     8 cm = _____ dm = _____ m

     b.   17 cm = _____ dm = _____ m

     c.   35 cm = _____ dm = _____ m

     d.   30 cm = _____ dm = _____ m

     e.  100 cm = _____ dm = _____ m

     f.  199 cm = _____ dm = _____ m

     g.  140 cm = _____ dm = _____ m

     h.  107 cm = _____ dm = _____ m

     i.  1078 cm = _____ dm = _____ m

     j.  1009 cm = _____ dm = _____ m

     k. 1500 cm = _____ dm = _____ m

     l.  1000 cm = _____ dm = _____ m

Here are the answers:

|       |             |            |       |             |             |
|-------|-------------|------------|-------|-------------|-------------|
| a.    | 0.8         | 0.08       | g.    | 14.0 = 14   | 1.40        |
| b.    | 1.7         | 0.17       | h.    | 10.7        | 1.07        |
| c.    | 3.5         | 0.35       | i.    | 107.8       | 10.78       |
| d.    | 3.0 = 3     | 0.30       | j.    | 100.9       | 10.09       |
| e.    | 10.0 = 10   | 1.00 = 1   | k.    | 150.0 = 150 | 15.00 = 15  |
| f.    | 19.9        | 1.99       | l.    | 100.0 = 100 | 10.00 = 10  |

  3.  Here are some slightly different ones.  Try them.

     a.    0.75 m = _____ cm    b. 0.  0.7 dm = _____ m

c.   0.8 dm = _____ cm          d.   0.09 m = _____ cm

e.   7.2 dm = _____ m           f.   6.7 dm = _____ cm

g.   1.02 m = _____ cm          h.   13.5 dm = _____ m

i.   17.4 dm = _____ cm         j.   5.4 m  = _____ cm

k. 107.4 dm = _____ m           l.  143.7 dm = _____ cm

Here are the answers:

a. 75                             b. 0.07
c. 8                              d. 9
e. 0.72                           f. 67
g. 102                            h. 1.35
i. 174                            j. 540
k. 10.74                          l. 1437

### TASK 4

To obtain experience in using the millimeter.

#### Materials

meter stick, paper decimeter with centimeter markings, adding machine tape

The *millimeter,* symbolized mm, is a more precise unit of measure than the centimeter, and is often used in industry.  A good example of the use of the millimeter is in photography; one of the most popular cameras is the 35-mm camera.  Metric tools often have a millimeter base for their calibration.  Can you think of other common objects that are described in millimeters?

#### Activity 1

Relating common objects to the millimeter.

1. a. Each centimeter contains 10 millimeters.  Mark your paper decimeter so that the millimeter markings show.  In the diagram in Figure 3.4, we have started this.

**Figure 3.4** Decimeter with millimeter markings.

   b. How many millimeters make a decimeter?

    c. How many millimeters make a meter? Check your answer
       with a meter stick marked in millimeters.

    d. Do you see why "milli" means "0.001" and "millimeter"
       means "0.001 of a meter"?

2. Measure the following to the closest millimeter.
    a. Your small fingernail
    b. Diameter of a penny, dime, or quarter
    c. Diameter of your ring, watch face, or buttons
    d. Length of a keyhole
    e. Width of match stick
    f. Diameter of screw or nail

### Activity 2
Estimating and verifying linear measures to the closest millimeter.

1. Estimate the segments shown in Figure 3.5 to the closest millimeter. Check by actually measuring.

(a)             (b)             (c)             (d)

**Figure 3.5** Estimate to the closest millimeter.

2. Estimate the lengths of these words on this page. Check by actually measuring.
    a. Check
    b. the
    c. of
    d. page
    e. measuring
    f. Activity

### Activity 3
Expressing millimeter measures in equivalent centimeter, decimeter, and meter forms.

Recall that:

$$10 \text{ mm} = 1 \text{ cm}$$
$$100 \text{ mm} = 1 \text{ dm}$$
$$1000 \text{ mm} = 1 \text{ m}$$

Therefore:

$$1 \text{ mm} = 0.1 \text{ cm} = 0.01 \text{ dm} = 0.001 \text{ m}$$

1. Use the information above to fill in the blanks.

    a.      8 mm = _____ cm = _____ dm = _____ m

    b.     13 mm = _____ cm = _____ dm = _____ m

    c.    186 mm = _____ cm = _____ dm = _____ m

    d.    105 mm = _____ cm = _____ dm = _____ m

    e.    998 mm = _____ cm = _____ dm = _____ m

    f.   1003 mm = _____ cm = _____ dm = _____ m

    g. 10004 mm = _____ cm = _____ dm = _____ m

    h. 20145 mm = _____ cm = _____ dm = _____ m

Here are the answers:

| | | | |
|---|---|---|---|
| a. | 0.8 | 0.08 | 0.008 |
| b. | 1.3 | 0.13 | 0.013 |
| c. | 18.6 | 1.86 | 0.186 |
| d. | 10.5 | 1.05 | 0.105 |
| e. | 99.8 | 9.98 | 0.998 |
| f. | 100.3 | 10.03 | 1.003 |
| g. | 1000.4 | 100.04 | 10.004 |
| h. | 2014.5 | 201.45 | 20.145 |

2. Here are some slightly different ones. Try them.

    a. 0.006 m = _____ mm        b. 0.075 m = _____ mm

    c. 0.106 m = _____ mm        d. 0.250 m = _____ mm

    e. 0.001 m = _____ mm        f. 0.010 m = _____ mm

    g. 0.100 m = _____ mm        h. 0.1   m = _____ mm

    i. 0.5   m = _____ mm        j. 0.65  m = _____ mm

Here are the answers

| | | | |
|---|---|---|---|
| a. | 6 | b. | 75 |
| c. | 106 | d. | 250 |
| e. | 1 | f. | 10 |
| g. | 100 | h. | 100 |
| i. | 500 | j. | 650 |

*TASK 5*

To obtain experience in using the dekameter, hectometer, and kilometer.

*Materials*

a ball of twine, scissors

1. With a partner, cut off a 10-meter length of twine. Mark each
meter length. This 10-meter length is called a *dekameter.*

2. With your dekameter twine, determine
   a. how far you can walk in 15 seconds when walking swiftly
   b. the combined length of five cars
   c. the combined length and width of this room

3. Tie together nine other dekameter twines to yours. This new
length is approximately 10 dekameters long and is called a *hectometer.*
How many meters make up a hectometer?

4. Name a place that you think is a hectometer from this room.
Check by pacing or by measuring with a meter tape.

5. If conditions permit, lay out a square hectometer outdoors.

6. If 10 hectometer twines were tied together, this new twine
length would be a *kilometer* long. How many meters make a kilometer? Name a place that is a kilometer from this room.

7. Do you see why "deka" means "ten," "hecto" means "a hundred," and "kilo" means "a thousand"?

## SUMMING UP

One of the advantages of the metric system is that there is a pattern
by which its units of measurements are formed. With linear measurements you have seen how all linear units were compared with the meter, the standard linear unit. These units and how they compare with
the meter are summarized below.

1 kilometer (km) = one thousand (1000) meters
1 hectometer (hm) = one hundred (100) meters
1 dekameter (dam) = ten (10) meters
1 meter (m) = one (1) meter
1 decimeter (dm) = one-tenth (0.1) of a meter
1 centimeter (cm) = one-hundredth (0.01) of a meter
1 millimeter (mm) = one-thousandth (0.001) of a meter

In each case the prefix of "meter" indicates how the units compare with the meter. These prefixes occur in the metric system many times and always have the same meaning.

kilo- means 1000
hecto- means 100
deka- means 10
deci- means 0.1
centi- means 0.01
milli- means 0.001

The most commonly used linear units are kilometer, meter, centimeter, and millimeter. The decimeter is increasingly being used in the elementary schools. The other units are rarely used.

## TIME OUT

### TASK 1
To experience and understand the approximate nature of measurement.

### Materials
meter stick

### Activity 1
1. Carefully measure the distance from the floor to the base of the doorknob to the closest millimeter.

2. Compare your result with results obtained by four other people. Did you all obtain the same measurements?

3. Measure the distance from the floor to the base of the switch plate to the closest millimeter.

4. Again, compare your result with those found by four other people. Did you *all* agree on the measurement?

5. If for each task, all five measurements did not agree, explain possible reasons for the difference.

If all five used the meter stick correctly (no careless human error), it is expected that the measures would differ (it may or may not be noticeable) because by its nature, measurement is inexact. Measurement in general terms consists of the following:

1. A unit is chosen,

2. The unit and the object to be measured are compared, and

3. The number of times the chosen unit is applied to the object is counted.

Since correct measurements differ due to the nature of measurement, we say measurements are only *approximate.*

### Activity 2

1. Measure the line in Figure 3.6 to the closest (a) mm, (b) cm, (c) dm.

---

**Figure 3.6** Line segment.

2. Which measurement is closest to the actual length of the line?

3. Which unit, the millimeter, centimeter, or decimeter, is the best one to use?

The smaller the unit of measure used, the more *precise* is the measurement. In the measurements above, the millimeter unit is the most precise and the decimeter is the least precise.

### TASK 2

To recognize the relationship of the metric prefixes and words in everyday life.

The prefixes of the metric system were chosen from Latin and Greek prefixes. Greek prefixes indicated multiples of the units and Latin prefixes denoted submultiples. Words other than those denoting metric units use these prefixes. Some of these words are listed below:

| | | |
|---|---|---|
| kilocycle | centennial | centipede |
| decile | hectograph | mile |
| millenium | decade | kilowatt |
| Decameron | decimal | December |

How many of these words do you recognize? Look up their meaning in a dictionary. Name other words using these Latin or Greek prefixes.

*EXERCISES*

1. What is the basic (standard) unit of linear measurement?

2. What do these prefixes mean: kilo-, centi-, and milli- ?

3. How do the kilometer, centimeter, and millimeter compare with the meter?

4. Arrange from smallest to largest: centimeter, millimeter, decimeter, hectometer, kilometer, meter.

5. Arrange the metric linear units from the most precise to the least precise.

6. What are the most commonly used metric linear units?

7. Using a linear unit of measurement, demonstrate the general nature of measurement. (See Chapter 2.)

8. Make a list of things that are currently described by linear metric units alone or along with customary units.

9. Research and report on the history of linear units of measurement, for example, the cubit, fathom, and furlong. (See Appendix II for references.)

10. Read and report on **About Measurement** by Margaret Friskey.

11. Make up a list of objects that children would enjoy measuring.

# chapter
# four

# APPLYING LINEAR
# UNITS: LENGTH,
# AREA, AND VOLUME

**LINEAR MEASURE**

*TASK 1*
To use linear units in various situations to determine length.

*Activity 1*
Determining perimeter.

Determine the perimeter of the objects named below to the closest unit indicated or preferred. (The perimeter is the boundary of an object.) First estimate, then accurately measure.

|   |  | *Estimate* | *Measurement* |
|---|---|---|---|
| 1. | The boundary of the playground | _____ | _____ |
| 2. | A sandbox | _____ | _____ |
| 3. | Any poster (centimeter) | _____ | _____ |
| 4. | Boundary formed by a jump rope (decimeter) | _____ | _____ |
| 5. | An identification card (centimeter) | _____ | _____ |
| 6. | This building | _____ | _____ |
| 7. | The shapes in Figure 4.1 (centimeter) | _____ | _____ |

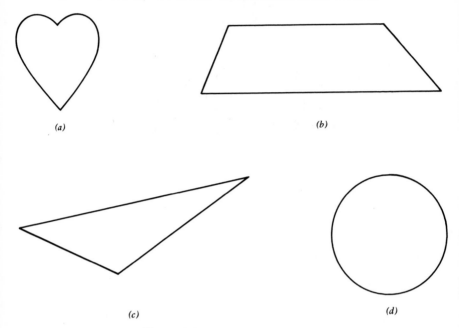

**Figure 4.1** Determining perimeter.

*Activity 2*
Making a vase.

*Materials*
yarn, string, or ribbon, large milk carton

Estimate the amount of yarn, string, or ribbon needed to wrap the milk carton, then actually wrap the milk carton.  How many meters of yarn, string or ribbon did you use?  What was the difference between your estimate and your actual measurement?  Use 0.5 m of yarn, string or ribbon (a different color if you wish) to make a design for the vase.  Make the design conform to specific metric lengths.

*Activity 3*
Draw the following:

1.  A triangle with sides 4 cm, 5 cm, and 6 cm.

2. A rectangle with sides 0.2 dm and 0.3 dm.

3. A triangle with sides 0.1 m, 0.10 m, and 0.100 m.

### Activity 4

Determine the following:

1. The size of your hat in centimeters (distance around head).

2. The size of your belt in centimeters (distance around waist).

3. The size of your ring in millimeters (distance around finger).

4. The size of your shoe in centimeters (foot length).

### Activity 5

Relating metric lengths to the environment.

1. Name objects in the environment that have the following approximate linear measurements:
   a. Five meters
   b. Twenty-five centimeters
   c. Thirty-five millimeters
   d. A hundred meters
   e. Ninety kilometers

2. Determine the approximate height from the ground of the following:
   a. A balance beam
   b. A lamp post
   c. A stop sign
   d. Three other objects found outdoors.

## AREA MEASURE

### TASK 1

To obtain experience in using the common area units.

#### Activity 1

#### Materials

cardboard or stiff paper, scissors

1. On the blackboard, draw a **square meter,** that is, a square region with side measures one meter long. The square meter is symbolized $m^2$.

2. On cardboard, draw a *square decimeter,* symbolized $dm^2$, and cut it out for later use.

3. On cardboard, draw a *square centimeter,* symbolized $cm^2$, and cut it out for later use.

4. Draw a *square millimeter,* symbolized $mm^2$. Do *not* cut out. Why?

### Activity 2
Making a comic collage

#### Materials
old comics and newspaper, scissors, glue or tape, colored pen, square decimeter, and square centimeter cardboards

1. Cut out a rectangular region with area 6 $dm^2$ from an old newspaper. Use this as a backing sheet.

2. Cut out from old comics regions with the following areas:

| | |
|---|---|
| a. 1 $cm^2$ | e. 30 $cm^2$ |
| b. 2 $cm^2$ | f. 20 $cm^2$ |
| c. 3 $dm^2$ | g. 20 $cm^2$ |
| d. 2 $dm^2$ | h. 10 $cm^2$ |

3. Make a design by attaching the comic cutouts to the backing sheet and by shading in with a colored pen regions with the following areas:

a. 50 $mm^2$        b. 10 $mm^2$        c. 25 $mm^2$

### Activity 3
Estimating and verifying area measurements.

1. To the closest square meter, estimate and measure the area of a parking space.

Estimate _____        Measurement _____

2. To the closest square decimeter, estimate and measure the area of the following:

| | *Estimate* | *Measurement* |
|---|---|---|
| a. A door mat | _____ | _____ |
| b. Your chair seat | _____ | _____ |
| c. The door | _____ | _____ |

|  | Estimate | Measurement |
|---|---|---|
| d. Your face | _____ | _____ |
| e. A placemat | _____ | _____ |
| f. An area rug | _____ | _____ |

3. To the closest square centimeter, estimate and measure the area of the following:

|  | | |
|---|---|---|
| a. All of your credit cards | _____ | _____ |
| b. An eraser top | _____ | _____ |
| c. The palm of your hand | _____ | _____ |
| d. A light switch plate | _____ | _____ |
| e. A mirror | _____ | _____ |
| f. A frying pan | _____ | _____ |

4. To the closest square millimeter, estimate and measure the area of the following:
   a. A postage stamp
   b. Your thumbnail
   c. Face of a watch
   d. A key

### Activity 4
Relating common area units to the environment.

1. Name objects in the environment that have the following approximate areas:
   a. A square centimeter
   b. Ten square centimeters
   c. A square decimeter
   d. Ten square decimeters
   e. A square meter
   f. Ten square meters
   g. Twenty-five square centimeters

2. Determine how much area is needed to play the game of hopscotch.

3. Determine the area of the following:

a. An average house lot
b. A stepping stone
c. The sidewalk area in front of your home

*Activity 5*
Being crafty.

*Materials*
florist wire, thread, pliers, scissors, oaktag or construction paper, meter tape or ruler

1. Make a mobile as follows:
   a. Using oaktag or construction paper make shapes that have the following areas:
      1 dm², 6 cm², 3 cm², 1000 mm², 100 mm², 10 cm², and 200 mm².
   b. Use thread to hang these shapes from the florist wire, forming a mobile. An example is shown in Figure 4.2.

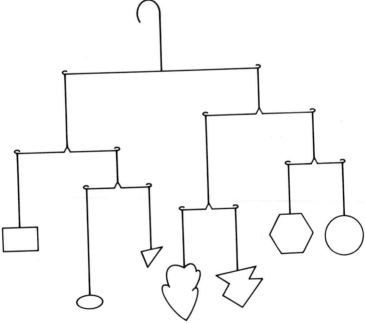

**Figure 4.2** A mobile.

How many centimeters of florist wire was used in the mobile?
How many centimeters of thread was used?

2. Make another mobile using area measurements of your choice.
Have someone estimate the area of each shape you make.

3. Make five or more gift cards or tags as follows:
   a. Obtain heavyweight paper or cardboard.
   b. Outline the shapes of your choice on the selected paper.
   c. Cut out the shapes.
   d. Use your imagination to design the cutout shape. An ex-
      ample is shown in Figure 4.3.

   Determine the area of each gift card or tag you made.

**Figure 4.3** A gift tag.

### *Activity 6*
Just imagine!

1. How many soda pop caps are needed to cover a square meter?

2. What area would be covered by all the loose change (pennies,
nickels, dimes, and quarters) of the people in this room?

3. What is the area of the sidewalk around a city block?

4. How many one-dollar bills are needed to "wallpaper" this
room?

5. How many textbooks are needed to cover a square dekameter?

6. How many cars can be parked in a square hectometer?

7. How many ants will occupy a space of one square decimeter?

8. How many pencil dots will cover an area of one square centimeter?

9. What is the area of your silhouette? Trace it and find out.

10. What is the area of a cylindrical tomato sauce can label if its

height is 15 cm and the circumference of the top of the can is 30 cm? Make the label to find out or to check your response.

The area unit that is the smallest is the most precise; the area unit that is the largest is the least precise. Hence, of the four area measures: square meter, square decimeter, square centimeter, and square millimeter, the square millimeter is the most precise and the square meter is the least precise. The unit of precision needed will depend on the task at hand.

### TASK 2
To discover the pattern of common area units.

#### Materials
square decimeter and square centimeter cardboards, drawing of a square meter or a poster the size of a square meter, metric ruler

1. Using a square decimeter cardboard and a square meter, determine how many square decimeters are in a square meter.

2. Using a square decimeter cardboard and a square centimeter cardboard, determine how many square centimeters are in a square decimeter.

3. Using a square centimeter and a metric ruler, determine, by drawing, the number of square millimeters in a square centimeter.

4. How many square millimeters are in a square decimeter?

5. How many square millimeters are in a square meter? Did you notice this pattern?

> There are 100 mm² in a square centimeter.
> There are 10 000 mm² in a square decimeter.
> There are 1 000 000 mm² in a square meter.

6. How many square centimeters are in a square meter?

### TASK 3
To use the pattern of the common units to determine equivalent area measures.

1. Draw a square decimeter. Shade in 33 square centimeters. What part of the square decimeter did you shade? Is 33 cm² equal to 0.33 dm²?

2. Draw a square centimeter. Shade in 45 square millimeters.

What part of the square centimeter did you shade? Is 45 mm² equal to 0.45 cm²?

3. Draw a square decimeter. Shade in 13 square millimeters. What part of the square decimeter did you shade? Is 13 mm² equal to 0.0013 dm²?

4. Draw rectangular regions with the following areas.

a. 0.3 dm²                    d. 0.90 cm²
b. 2.5 dm²                    e. 0.01 m²
c. 0.15 dm²

5. Name common objects with the following approximate area:

a. 30.5 cm²
b. 1.7 cm²
c. 1.75 m²

6. Use decimal fractions in expressing equivalent area measures by completing the following statements. The first one is done for you.

a. 93 cm² = __0.93__ dm²

b. 91 mm² = _____ cm²

c. 89 dm² = _____ m²

d. 45 cm² = _____ dm²

e. 23 mm² = _____ cm² = _____ dm² = _____ m²

f. 30 mm² = _____ cm² = _____ dm² = _____ m²

Are your answers reasonable relative to actual sizes? Here are the answers:

a. 0.91
b. 0.89
d. 0.45
e. 0.23; 0.0023; 0.000 023
f. 0.30; 0.0030; 0.000 030


# VOLUME MEASURE

*TASK 1*

To obtain experience in using the common volume units.

*Activity 1*

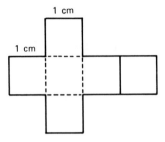

**Figure 4.4** Cube pattern.

**Figure 4.5** One cubic centimeter.

### Materials
cardboard, construction or stiff paper, scissors, tape (about 1 cm wide)

1.  On your cardboard draw the pattern shown in Figure 4.4. (Note that the pattern consists of six squares, each with side measure of one centimeter.)

2.  Cut out the pattern along the solid lines.

3.  Fold inward along the dotted lines.

4.  Tape in place to form a cube. (See Figure 4.5.)
    Since the sides of the squares measure one centimeter and since the object formed is a cube, we say the object has volume measure of *one cubic centimeter,* symbolized: $cm^3$.

5.  Approximately how many grains of rice will fit into the cubic centimeter? Find out.

6.  Approximately how many cubic centimeters will fill a shoe box? A closet?

7.  Approximately what volume is needed to hold a dozen cherries?

(not actual size)

**Figure 4.6** Cube pattern without a lid.

## Activity 2

### Materials

cardboard, construction or stiff paper, scissors, tape

1. Repeat Steps 1 through 4 in Activity 1 above, but use only five squares, and the side of each square should measure one decimeter. See Figure 4.6.

2. The object formed is a cube without a lid and has volume measure of **one cubic decimeter,** symbolized: dm³.

3. How many cubic centimeters will fit into your cubic decimeter? Find out.

4. About how many whole pieces of chalk will fit into a square decimeter?

5. Approximately how many cubic decimeters will fill a bathtub?

6. About what volume is needed to hold 12 large grapefruits?

### Activity 3

Relating the cubic centimeter and cubic decimeter to Cuisenaire rods.

### Materials

Cuisenaire rods

1. Give the volume of each of the Cuisenaire rods. What is the combined volume of these rods?

2. Use Cuisenaire rods to build imaginative structures with the following volumes.

     a. 19 cm³
     b. 26 cm³
     c. 32 cm³
     d. 57 cm³
     e. 65 cm³

3. How many orange rods are needed to form one layer of rods in the cubic decimeter you made in Activity 2? What is the volume of this layer? How many of these layers are needed to fill your cubic decimeter?

### Activity 4

**Materials**
12 meter sticks, tape

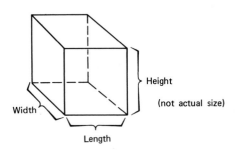

**Figure 4.7**  Cubic meter.

1. Use 12 meter sticks to make a *cubic meter,* symbolized: $m^3$. See Figure 4.7.

2. What is the height of the cubic meter? What is its length? What is its width?

3. What is the area of each face of the cubic meter?

4. How many people can be placed in the cubic meter? Find out.

5. About how many basketballs are needed to fill a cubic meter?

6. About what volume is needed to hold a thousand large pineapples?

7. About how many cubic meters will this room hold?

### Activity 5
Estimating and verifying volume measures.

**Materials**
four different sized boxes, common cartons

1. To the closest cubic centimeter or cubic decimeter, estimate the volume of boxes A to D. Now measure the volume of these boxes.

|  | *Estimate* | *Measurement* |
|---|---|---|
| Box A | ——— | ——— |
| Box B | ——— | ——— |
| Box C | ——— | ——— |
| Box D | ——— | ——— |

2. To the closest cubic centimeter, cubic decimeter, or cubic meter, estimate and then measure the volume of the following common objects:

|  | *Estimate* | *Measurement* |
|---|---|---|
| a. matchbox | ——— | ——— |
| b. cigarette case | ——— | ——— |
| c. chalk box | ——— | ——— |
| d. ice cream carton | ——— | ——— |
| e. tissue carton | ——— | ——— |
| f. this classroom | ——— | ——— |

### *Activity 6*

Relating common volume units to the environment. Name objects in the environment that have the following approximate volume:

1. A cubic meter
2. Two cubic meters
3. Ten cubic meters
4. A hundred cubic meters
5. A cubic centimeter
6. Ten cubic centimeters
7. A hundred cubic centimeters
8. A cubic decimeter

9. Ten cubic decimeters

10. A hundred cubic decimeters

### Activity 7
Meeting metrics head on.

### Materials
empty boxes of different sizes, glue or tape

1. Use boxes of different sizes to construct the heads of members of the "metric family." For example, Mr. Metric can be built from a large cereal box (face), two small cereal boxes (ears) and a matchbox (nose). An example is shown in Figure 4.8.

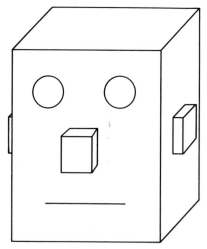

**Figure 4.8** "Metric family" member.

2. Estimate and then measure the volumes of the heads.

3. Determine the surface areas (the area that shows) of the heads.

4. Which member of the family has the biggest head (volume)? The smallest head?

5. Arrange the heads from large to small relative to surface area. Is this ordering the same relative to volume?

### TASK 2
To discover the volume relationship among the cubic centimeter, cubic decimeter, and cubic meter.

### Materials
your cubic centimeter, cubic decimeter, and cubic meter, additional cubic centimeters (classmate's or commercial ones)

1.  Use cubic centimeters and a cubic decimeter to estimate how many cubic centimeters will fit into the cubic decimeter. Use additional cubic centimeters and verify your estimate.

2.  How many cubic decimeters will form the bottom layer of a cubic meter? How many layers can be formed? How many cubic decimeters in a cubic meter?

3.  How many cubic centimeters will fit into a cubic meter? Did you get these answers?
1.  1000
2.  100; 10; 1000
3.  1 000 000

### TASK 3
To realize the volume relationships among the cubic centimeter, cubic decimeter, and cubic meter.

### Materials
a cubic meter, cubic decimeters, and cubic centimeters

1.  Place three cubic centimeters into a cubic decimeter. What part of the cubic decimeter is filled? Is $3 \text{ cm}^3 = 0.003 \text{ dm}^3$?

2.  Place four cubic decimeters into the cubic meter. What part of the cubic meter is filled? Is $4 \text{ dm}^3 = 0.004 \text{ m}^3$?

3.  Identify objects in the environment that have the following approximate volumes:
    a. $2.85 \text{ dm}^3$          b. $0.300 \text{ m}^3$          c. $0.030 \text{ m}^3$

4.  Complete the following statements. Check to see if your answers are reasonable relative to actual volume sizes.

    a.  $93 \text{ cm}^3 = $ _____ $\text{dm}^3$     d.  $199 \text{ dm}^3 = $ _____ $\text{m}^3$

    b.  $135 \text{ cm}^3 = $ _____ $\text{dm}^3$     e.  $500 \text{ mm}^3 = $ _____ $\text{cm}^3$

    c.  $12 \text{ dm}^3 = $ _____ $\text{m}^3$     f.  $0.001 \text{ m}^3 = $ _____ $\text{mm}^3$

Here are the answers:
a. 0.093             d. 0.199
b. 0.135             e. 0.500
c. 0.012             f. 1 000 000

## *EXERCISES*

1. What are the side measurements of a square meter? a square decimeter? a square centimeter?

2. Suggest three common objects you could always associate with the square meter, square decimeter, and square centimeter.

3. How do the square meter, square decimeter, and square centimeter compare in area?

4. What is the area of each of the faces of a cubic centimeter? of a cubic decimeter? of a cubic meter?

5. Suggest three common objects you could always associate with the cubic meter, cubic decimeter, and cubic centimeter.

6. How do the cubic meter, cubic decimeter, and cubic centimeter compare in volume?

7. Give the symbol for each of the following: a centimeter, a square centimeter, a cubic centimeter, a meter, a square meter, and a cubic meter.

8. Try out the length activity for the elementary classroom in Appendix III.

9. Try out the area activity for the elementary classroom in Appendix III.

10. Design an activity for elementary school children that gives them experience in using square centimeter and square decimeter units.

11. Design an activity for elementary school children to provide experience in using cubic centimeter and cubic decimeter units.

12. Bring to class several interesting objects to measure for volume. Have someone in class measure them. Check the measurements.

# chapter five

# CAPACITY

To obtain experience in using the standard units of capacity.

As the meter is the basic unit of length, so the *liter,* symbolized $\ell$, is the basic unit of capacity (the amount a container could hold) in the metric system. The liter is defined as a cubic decimeter (0.001 m³). It is generally used when measuring the volume of liquids.

1 kiloliter (k$\ell$)   = one thousand (1000) liters
1 hectoliter (h$\ell$) = one hundred (100) liters
1 dekaliter (da$\ell$) = ten (10) liters
1 liter ($\ell$)       = one (1) liter
1 deciliter (d$\ell$)  = one-tenth (0.1) of a liter
1 centiliter (c$\ell$)  = one-hundredth (0.01) of a liter
1 milliliter (m$\ell$) = one-thousandth (0.001) of a liter

Common objects with the above capacities (approximate) are:

1 kiloliter (k$\ell$)    a good-sized refrigerator
1 hectoliter (h$\ell$)  a large aquarium
1 dekaliter (da$\ell$)  a pail
1 liter ($\ell$)         a can of automobile oil
1 deciliter (d$\ell$)   a half cup of coffee
1 centiliter (c$\ell$)  2 teaspoons
1 milliliter (m$\ell$) 10 drops of water

### Activity 1
With your group members, make six standard-capacity units: 1 $\ell$, 500 m$\ell$, 250 m$\ell$, 100 m$\ell$ = 1 d$\ell$, 50 m$\ell$, 10 m$\ell$ = 1 c$\ell$.

### Materials
liter, 500-m$\ell$, 250-m$\ell$, 100-m$\ell$, 50-m$\ell$, and 10-m$\ell$ beakers; two large, two medium, one small plastic or paper cup; a container of water (rice or sand may be substituted); masking tape; a large mayonnaise, jam, or coffee jar

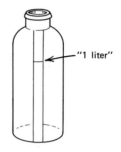

**Figure 5.1** Making a liter container.

1. Put a strip of masking tape vertically down one side of your large jar.

2. Measure a liter of water into the liter beaker, then pour it into your jar.

3. Mark the water level on the masking tape and label it "1 liter" (as shown in Figure 5.1.

4. Similarly, using the five plastic or paper cups and the classroom beakers, make containers of 500 mℓ, 250 mℓ, 100 mℓ, 50 mℓ, and 10 mℓ.

### Activity 2
Work with your group members to graduate the set of standard capacity units you made.

### Materials
your set of nongraduated standard-capacity containers; a container of water (rice or sand may be substituted)

1. a. Using your 500-mℓ container, measure 500 mℓ of water into your liter.
   b. Mark the water level on the masking tape and label it "500 mℓ."
   c. Measure another 500 mℓ of water into your liter. Is the water level at the liter mark?
   d. Return the water to its original container.

2. a. Using your 250-mℓ container, measure 250 mℓ of water into your liter container. Mark and label each 250 mℓ level on the masking tape until you reach the liter mark.

b. Fill in the blanks.

(1) $\frac{1}{4}$ liter = _____ mℓ = __0.25__ ℓ

(2) $\frac{1}{2}$ liter = _____ mℓ = _____ ℓ

(3) $\frac{3}{4}$ liter = _____ mℓ = _____ ℓ

(4) 1 liter = _____ mℓ = _____ ℓ

c. Return the water to its original container.

3. a. Using your 100-mℓ container, measure 100 mℓ of water into your 500-mℓ container. Mark and label each 100-mℓ level on the masking tape until you reach the 500-mℓ mark.

b. Fill in the blanks.

(1) $\frac{1}{10}$ liter = _____ mℓ = __0.1__ ℓ

(2) $\frac{2}{10}$ liter = _____ mℓ = _____ ℓ

(3) $\frac{3}{10}$ liter = _____ mℓ = _____ ℓ

(4) $\frac{4}{10}$ liter = _____ mℓ = _____ ℓ

4. a. Using your 50-mℓ container, measure 50 mℓ of water into your 100-mℓ container. Mark and label each 50-mℓ level on the masking tape until you reach the 100-mℓ mark.

b. Return the water into its original container.

5. a. Using your 50-mℓ container, measure 50 mℓ of water into your 250-mℓ container. Mark and label each 50 mℓ level on the masking tape until you reach the 250 mℓ mark.

b. Return the water into its original container.

6. a. Using your 10-mℓ container, measure 10 mℓ of water into your 50-mℓ container. Mark and label each 10-mℓ level on the masking tape until you reach the 50-mℓ mark.

Save your set of containers for future use.

Table 5.1  Capacity of Common Items

| Container Number | Estimate | Measurement | Difference |
|---|---|---|---|
| 1 | | | |
| 2 | | | |
| 3 | | | |
| 4 | | | |
| 5 | | | |

*Activity 3*
Estimating and verifying the capacity of common items.

*Materials*
your set of standard-capacity units, numbered common containers of various sizes (e.g., Coke bottle, milk carton), a container of water (rice or sand may be substituted)

1.  Estimate the capacities of the numbered containers.
2.  Record your estimates in Table 5.1 above.
3.  By measuring, determine the actual capacity of each container.
4.  Calculate the difference between your estimate and your actual measures.  Record.
5.  Compare your accuracy in estimating capacity with someone.

*Activity 4*
Estimate the capacities of the following items:

1.  A roasting pan
2.  A standard sized bathtub
3.  A tea kettle
4.  The kitchen sink in your house
5.  An average sized washing machine

Verify your estimates when possible.

*Activity 5*
Name at least three objects in your home and/or community which have items with the following approximate capacities.

1. Two liters
2. Five milliliters
3. Five hundred milliliters
4. Ten liters
5. Two hundred fifty milliliters
6. One liter
7. One kiloliter

### TASK 2
To verify the pattern of the standard-capacity units.

#### Materials
a centiliter (0.01 liter), a deciliter (0.1 liter), a liter, and a dekaliter (10 liters) container; a source of water or a bucketful of water (rice or sand may be substituted)

1. By pouring 10 centiliters of water into your deciliter container, verify that 10 centiliters = 1 deciliter.

2. By pouring 10 deciliters of water into your liter container, verify that 10 deciliters = 1 liter.

3. By pouring 10 liters of water into your dekaliter container verify that 10 liters = 1 dekaliter.

### TASK 3
To use the pattern of the capacity units to determine equivalent capacities of common objects.

Use the pattern of the standard capacity units and the capacities of common objects to answer the following questions:
   a. About how many 250-mℓ containers of sand does it take to fill a large coffee can?
   b. About how many bucketfuls of water does it take to fill a kiloliter?
   c. About how many cans of automobile oil does it take to fill a 10-liter container?
   d. About how many drops of water will a 2-mℓ eyedropper hold?
   e. About how many liters of water are needed to fill a large aquarium?

f. About how many cups of coffee will a 3-liter coffee percolator make?

Did you get these answers?
a. About 8 containers (#2 size can)
b. About 100 bucketfuls of water
c. About 10 cans
d. About 20 drops of water
e. About 100 liters
f. About 15 cups

### TASK 4
To use decimal fractions in expressing equivalent capacities.

The most commonly used capacity units are the liter and milliliter. For this reason we need to be able to express milliliter units in liter units and vice versa with some facility. This facility is based on an understanding of decimal fractions (see page 5 for a quick review), usage, and on the knowledge that 1000 milliliters equals one liter.

Test your facility in expressing capacity in equivalent liter and milliliter units by completing the following statements:

1. $1 \text{ m}\ell = $ _____ $\ell$ 
2. $0.01 \ \ell = $ _____ $\text{m}\ell$

3. $100 \text{ m}\ell = $ _____ $\ell$ 
4. $3 \text{ m}\ell = $ _____ $\ell$

5. _____ $\ell = 40 \text{ m}\ell$ 
6. _____ $\text{m}\ell = 0.50 \ \ell$

7. $7.10 \ \ell = $ _____ $\text{m}\ell$ 
8. $2400 \text{ m}\ell = $ _____ $\ell$

9. $53.8 \ \ell = $ _____ $\text{m}\ell$ 
10. $0.05 \ \ell = $ _____ $\text{m}\ell$

11. $0.350 \ \ell = $ _____ $\text{m}\ell$ 
12. $0.100 \ \ell = $ _____ $\text{m}\ell$

Here are the answers:
1. 0.001
2. 10
3. 0.1
4. 0.003
5. 0.04
6. 500
7. 7100
8. 2.400
9. 53 800
10. 50
11. 350
12. 100

### EXERCISES
1. What is the basic unit of capacity?
2. Identify a common object you could associate with a liter capacity.

3. What are the most commonly used units of capacity?
4. How does the liter compare in capacity with the quart?
5. Make a list of things that are commonly described in milliliters.
6. What is the difference in capacity between a cubic centimeter and a milliliter?
7. What is the difference in capacity between a cubic decimeter and a liter?
8. Research and report on the history of the units of measure for capacity.
9. Using a unit of measure for capacity, demonstrate the general nature of measurement. (See Chapter 2.)
10. How many milliliters of water do you think a metric cup should hold?
11. Present the capacity activity in Appendix III in an elementary classroom.

# chapter
# six

# MASS

### TASK 1
To construct a simple balance.

#### Materials
two whipped cream containers or butter tubs; one wire clothes hanger; three pieces of twine or thread, each about 30 cm long

1.  Punch a hole about one centimeter down from the top edge of each container.

2.  In each of the tubs punch another hole directly across from the first hole.

3.  Tie one end of the twine to one of the holes.

4.  Loop the twine through the clothes hanger and then tie the other end to the other hole in the container.

5.  Repeat Steps 3 and 4 for the other container. Be sure the containers hang down the same distance. Spread the containers to opposite ends of the clothes hanger.

6.  Tie the third piece of twine to the hook of the clothes hanger.

7.  Check for *balance* by hanging the device by the twine. Balance occurs when the bottom wires hang horizontally. Adjust the containers until balance occurs (see Figure 6.1).

*Mass* is the *amount of material* in an object. If two objects have the same mass, they will balance one another. Your simple balance will help you determine equal masses. In everyday life the word "weight" is usually used as a synonym for the word "mass." Technically, these terms describe different concepts. Weight is a force; it expresses the pull of gravity on objects. Astronauts weigh less on the moon than on earth, but their masses are the same at both places.

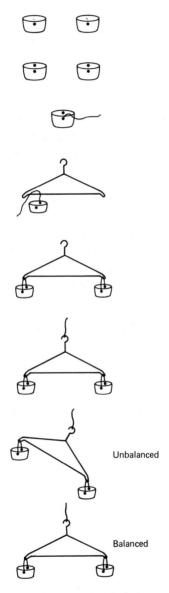

**Figure 6.1** Making a simple balance.

### TASK 2
To make and use arbitrary units of mass.

#### Materials
clothes hanger balance, Play-Doh, elbow macaroni, bottle caps, paper clips

1. Use the Play-Doh to make a round ball about 1.5 cm in diameter. This ball will be your *mass unit* for Task 2. Use your clothes hanger balance to answer the following questions:
   a. How many pieces of macaroni are needed to balance your mass unit?
   b. How many paper clips are needed to balance your mass unit?
   c. Which has greater mass, a paper clip or a piece of macaroni?
2. Make additional balls of Play-Doh and use your balance to find the following:
   a. Another round ball with the same mass as your mass unit.
   b. Two round balls each with a mass of one-half of your mass unit.
   c. Two round balls each with a mass of twice that of your mass unit.
   d. A round ball with three times the mass of your mass unit.
   e. A round ball with five times the mass of your mass unit.
   f. A round ball with ten times the mass of your mass unit.
3. Use your balance and determine, in terms of your mass unit, the mass of the following objects:
   a. An eraser
   b. A handful of paper clips
   c. Pieces of chalk
   d. Varying amounts of bottle caps
   e. Any three items of your choice
4. Compare your answers to number 3 above with another person. Do your answers agree?

### TASK 3
To obtain experience in using the standard units of mass.

As the meter is the basic unit of length and the liter is the basic unit of capacity, so the *kilogram* is the basic unit of mass in the metric system. The kilogram is defined as the mass of the International Prototype Kilogram, a cylinder of platinum-iridium alloy kept by the International Bureau of Weights and Measures at Sevres, France. The kilogram has the mass of one cubic decimeter of water at the temperature of maximum density. The other mass units are compared with the kilogram as follows:

1 metric ton (t)   = one million (1 000 000) grams
1 kilogram (kg)    = one thousand (1000) grams
1 hectogram (hg) = one hundred (100) grams
1 dekagram (dag) = ten (10) grams
1 gram (g)          = one (1) gram
1 decigram (dg)   = one-tenth (0.1) of a gram
1 centigram (cg)  = one-hundredth (0.01) of a gram
1 milligram (mg)  = one-thousandth (0.001) of a gram

Common objects with the above mass units (approximately) are:

| | |
|---|---|
| a Volkswagen Beetle | 1 metric ton (t) |
| a pineapple | 1 kilogram (kg) |
| a flashlight battery, size D | 1 hectogram (hg) |
| a wooden clothespin | 1 dekagram (dag) |
| 2 paper clips | 1 gram (g) |
| a straight pin | 1 decigram (dg) |
| a grain of raw rice | 1 centigram (cg) |
| a grain of salt | 1 milligram (mg) |

With the exception of the Volkswagen Beetle and the grain of salt, the objects named above (or those with equal masses) are on display. Beside each object is its approximate mass unit. Lift each object and its corresponding mass unit to develop a feel for the standard mass units and common representatives of those mass units.

### Activity 1
Estimating and verifying the mass of objects.

### Materials
a balance and a set of weights, 10 bottle caps, 10 classroom objects, pencil or pen

1. List the classroom objects whose mass you plan to estimate. Use the table given on the next page.

2. Estimate the approximate mass of each object by lifting it and record your estimates.

Table **6.1** Determining the Mass of Objects

| Object | Estimate | Measurement | Difference |
| --- | --- | --- | --- |
|  |  |  |  |

3. By using a balance, find and record the mass of each object.

4. Determine the difference between your estimation and the measured mass for each listed object.

If you wish to continue this task of estimating and verifying objects, estimate and verify the mass of the following:

|  | *Estimate* | *Measurement* |
| --- | --- | --- |
| 10 bottle caps | _____ | _____ |
| a milk carton | _____ | _____ |
| a small plastic carton | _____ | _____ |
| a large plastic cup | _____ | _____ |
| balls of Play-Doh | _____ | _____ |
| three pieces of dry macaroni | _____ | _____ |
| other objects of your choice | _____ | _____ |

### *TASK 4*
To verify the pattern of the mass units.

#### *Materials*
10 1-gram weights, 10 1-dekagram weights, 10 1-hectogram weights, 1 1-kilogram weight

If two sets of objects balance, they have equal masses.  By using a balance, verify that each of the following pairs of sets have equal masses:

| Set A | Set B |
|---|---|
| 1.  10 1-g weights | 1 dekagram weight |
| 2.  10 1-dag weights | 1 hectogram weight |
| 3.  10 1-hg weights | 1 kilogram weight |
| 4.  10 1-g weights, 9 1-dag weights | 1 hectogram weight |
| 5.   9 1-hg weights, 10 1-dag weights | 1 kilogram weight |

### TASK 5
To use the pattern of the mass units to determine equivalent masses of common objects.

#### Materials
the pattern of the mass units as found on page 67, the mass of common objects as found on page 67.

1.  Use the pattern of the mass units and the given masses of common objects to answer the following questions.

    a. About how many straight pins does it take to have a mass of 1 gram?  of 1 kilogram?

    b. About how many size D batteries does it take to have the mass of a pineapple?

    c. About how many pineapples does it take to have the mass of a Volkswagen Beetle?

    d. About how many paper clips does it take to have the mass of a size D battery?

    e. About how many clothespins does it take to have the mass of a kilogram?

    f. About how many grains of raw rice does it take to have the mass of a kilogram?

    g. About how many grains of salt does it take to have the mass of 1 gram?  of 1 kilogram?

    h. About how many paper clips does it take to have the mass of a clothespin?

2.  Verify your answers whenever possible.

Here are the answers for number 1.
a. About 10 pins; about 10 000 pins
b. About 10 batteries
c. About 1000 pineapples
d. About 200 paper clips
e. About 100 clothespins
f. About 100 000 grains
g. About 1000 grains; about 1 000 000 grains
h. About 20 paper clips

### TASK 6

To use decimal fractions in expressing equivalent masses.

The most commonly used mass units are the gram and the kilogram. For this reason we need to be able to express gram units in kilogram units and vice versa with some facility. This facility is based on an understanding of decimal fractions, usage, and the knowledge that 1000 grams equals 1 kilogram.

Test your facility in expressing mass in equivalent gram and kilogram units by completing the following statements:

1. 1 g = _____ kg    2. 0.01 kg = _____ g

3. 100 g = _____ kg    4. 3 g = _____ kg

5. _____ kg = 40 g    6. _____ g= 0.500 kg

7. 7.10 kg = _____ g    8. 2 400 g = _____ kg

9. 53.8 kg = _____ g    10. 0.05 kg = _____ g

11. 0.350 kg = _____ g    12. 0.100 kg = _____ g

Did you get these answers?
1. 0.001
3. 0.100
5. 0.040
7. 7100
9. 53 800
11. 350

2. 10
4. 0.003
6. 500
8. 2400
10. 50
12. 100

### EXERCISES

1. What is mass?
2. What is the basic unit of mass?
3. What are the most commonly used units of mass?

4. How does the gram compare in mass with the kilogram?
5. Identify a few common objects you could associate with the mass of a gram and with the mass of a kilogram.
6. Research and report on the history of units of mass.
7. What are arbitrary units of mass?
8. How much would a "pound cake" weigh in gram units?
9. Demonstrate and explain the nature of measurement using arbitrary units of mass.
10. Investigate and report to the class different and inexpensive ways to make a simple balance.
11. Try out the mass activity in Appendix III in an elementary classroom.

# chapter
# *seven*

# TEMPERATURE

### TASK 1
To obtain experience in using the Celsius thermometer.

#### Materials
Celsius thermometer (the clinical thermometer is unsafe to use), hot and cold water, salt, ice, boiling water or hot coffee or tea, aquarium or terrarium (optional)

The Celsius thermometer is consistent with the other metric measurements; that is, the thermometer is decimally based. The thermometer is divided into a hundred equal parts between the freezing point and the boiling point of water. The point at which water freezes is $0°$ and the point at which water boils at a pressure of one atmosphere is $100°$. (See Figure 7.1.)

1. Check your body temperature by placing the thermometer at your elbow. Hold the thermometer in your closed forearm for about three minutes (see Figure 7.2). Record the result in Table 7.1 on the following page.

*2. Let the hot-water tap run until the water is hot. Take a container of hot water and determine its temperature by placing the thermometer in it for 2 minutes. Record your result.

3. Repeat step 2 above using cold water.

4. Check the room temperature by leaving the thermometer out on the table top. Record the result.

5. Pack your thermometer in melting ice for three minutes. Check and record the temperature. Mix some salt with the ice. Now check and record the temperature.

*6. Boil some water. Let it boil for one minute. Check the tem-

*Hot coffee or tea might be substituted.

72

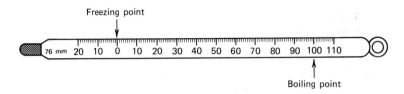

Figure 7.1  The Celsius thermometer.

Figure 7.2  Taking body temperature

Table 7.1  Temperature of Common Objects

|  | Readings | | |
| Elements | Self | Neighbor 1 | Neighbor 2 |
| --- | --- | --- | --- |
| 1.  Elbow | | | |
| 2.  Hot water | | | |
| 3.  Cold water | | | |
| 4.  Room | | | |
| 5.  Ice | | | |
| 6.  Ice and salt | | | |
| 7.  Boiling water | | | |
| 8.  Aquarium or terrarium | | | |

perature of the water at the top of the container and then at the bottom.  Record your result.

7.  Check the water temperature of an aquarium or check the temperature of a terrarium.  Record your result.

### TASK 2
To recognize and explain difference in temperature readings.

1. Obtain the results of Task 1 above from two of your neighbors. Record them in your table and compare the three sets of results. Were all three readings identical?

2. If your readings were not identical, explain why not.

### TASK 3
To apply knowledge of the Celsius thermometer.

1. Above what temperature would a child have a fever?

2. If your plumber set your hot water heater at 90°C, would the water be warm, normal, or hot?

3. At what temperature will ice melt?

4. What is the average temperature in your state in January? in July?

5. On a cold winter evening at what temperature would you set the thermostat?

6. What would be the temperature on a warm, balmy day?

7. In your town, what is the difference between the temperature at 8 o'clock in the morning and at noon?

8. When would it be too cold to travel without gloves?

9. At what temperature do you like your bath water?

10. At what temperature would you set the oven to bake a cake?

## SUMMING UP

### TASK
To discover and experience the relationships of mass, capacity, and linear measures.

One advantage of the metric system is that the same pattern of prefixes is used for mass, capacity, and linear measures. We review these prefixes here:

|  |  |
|---|---|
| kilo- | one thousand (1000) |
| hecto- | one hundred (100) |
| deka- | ten (10) |

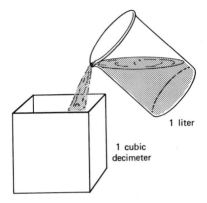

1 liter

1 cubic
decimeter

**Figure 7.3** Relationship between volume and capacity.

| deci- | one-tenth (0.1) |
| centi- | one-hundredth (0.01) |
| milli- | one-thousandth (0.001) |

Another advantage of the metric system is the simple relationship between capacity and volume measurements. The following activity will give you a "first-hand" experience with this relationship.

### Activity 1

**Materials**
a plastic cubic decimeter, a liter of water

1. Pour the liter of water into the plastic cubic decimeter (see Figure 7.3).
   - a. Did you show that a cubic decimeter equals a liter?
   - b. If your result was a bit inaccurate, explain why.
2. a. What is the volume of the cubic decimeter in terms of cubic centimeters?
   - b. How many milliliters of water can you pour into a cubic decimeter?
   - c. Did you show that 1000 mℓ = 1000 cubic centimeters?
3. If 1000 mℓ = 1000 cubic centimeters, then 1 mℓ = 1 cubic centimeter. Did you show that 1 mℓ = 1 cubic centimeter?

Still another advantage of the metric system is the relationship between the mass and capacity of pure water at 4°C. Activity 2 will help you see this relationship.

**Table 7.2** The Relationship Between Capacity and Mass

| Mass of Beaker | Mass of Beaker + Mass of Water | Mass of Water |
|:---:|:---:|:---:|
| (#1) | (#3) | (answer #3- answer #1) |

### Activity 2

**Materials**

your 1000-mℓ or liter beaker, 1000 mℓ or liter of distilled water at 4°C (tap water may be substituted), balance with weights

1. Find the mass of your empty 1000-mℓ or liter beaker. Record your findings in the table given.

2. Pour 1000 mℓ or liter of water into your beaker.

3. Find the mass of the beaker of water. Record your result in the table.

4. a. Did you show that 1000 mℓ or a liter of water = 1000 g of water? (See Figure 7.4.)

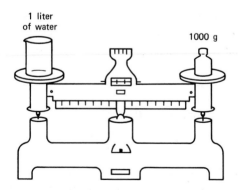

**Figure 7.4** Relationship between capacity and mass.

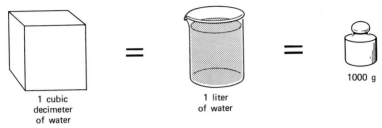

1 cubic
decimeter
of water

1 liter
of water

1000 g

**Figure 7.5** The relationship of volume, capacity, and mass.

b. If your results were a bit inaccurate, explain why.

5. Did you show that 1 ml of water = 1 g of water?

Based on the results of Activities 1 and 2, do you see how volume, capacity, and mass are related? When using distilled water at 4°C, a cubic decimeter (1000 cubic centimeters) has capacity of 1 liter (1000 ml) and mass of 1000 grams. (See Figure 7.5.)

### Exercises

1. Research and report on the history of the development of the Fahrenheit and Celsius thermometers.

2. Read about and demonstrate to the class how to make a simple nonstandard thermometer.

3. Using the Celsius scale, list those situations and temperatures with which everyone should be familiar. For example: boiling point of water 100°C.

4. (a) Give the meaning of these prefixes: kilo-, hecto-, deka-, deci-, centi-, milli-.
   (b) How are these prefixes related to one another?

5. In the metric system, what is the relationship between volume and capacity? Is there a similar relationship between volume and capacity in the English customary system?

6. In the metric system how are mass, volume, and capacity related? Is there a similar relationship in the English customary system?

7. Review three elementary textbook series and report to the class on how they introduce temperature concepts.

8. Try out the sample classroom activity that uses Kanten (Japanese gelatin) in Appendix III.

## TIME OUT

### *TASK*
To assess your knowledge of metric units.

Below is a self-test. Use it to determine your knowledge of the metric units studied.

Metric Self-Test

Part A

1. Estimate:
   a. The length of this page.
   b. The area of this page.
   c. The volume of an individual cereal box.
   d. The capacity of a small coffee can.

2. Name objects in your immediate environment that have the following approximate measurements.
   a. 10 meters
   b. 100 square centimeters
   c. 250 cubic centimeters
   d. 0.5 kilogram
   e. 0.5 liter

3. Check your answers to 1 and 2 above by actually measuring.

Part B

1. Write these measures in abbreviated form using the meter as the unit of length.

   a. 16 dm = _____ m

   b. Seven and five-tenths meters = _____ m

2. Fill in the blanks.

   a. 23 cm = _____ dm = _____ m

   b. 1480 cm = _____ dm = _____ m

   c. 0.87 m = _____ cm

   d. 12.5 dm = _____ m

   e. 72.8 m = _____ cm

3. Fill in the blanks.

    a. 46 mm = _____ cm = _____ dm = _____ m

    b. 20 230 mm = _____ cm = _____ dm = _____ m

    c. 4.217 m = _____ mm

    d. 0.003 m = _____ mm

4. Answer the following:

    a. A dekameter is equivalent to _____ decimeters.

    b. A kilometer is equivalent to _____ meters.

5. Determine the perimeter of a rectangle with sides 0.3 dm and 0.5 dm.

6. Complete the following statements:

    a. $72 \text{ dm}^2$ = _____ $\text{m}^2$

    b. $95 \text{ cm}^2$ = _____ $\text{dm}^2$

    c. $46 \text{ mm}^2$ = _____ $\text{cm}^2$ = _____ $\text{dm}^2$ = _____ $\text{m}^2$

7. Draw a cube that has a volume measure of one cubic centimeter.

8. Complete the following statements:

    a. 0.04 ℓ = _____ mℓ

    b. _____ ℓ = 4200 mℓ

    c. 82.8 ℓ = _____ mℓ

9. Fill in the blanks:

    a. 4250 g = _____ kg

    b. 0.210 kg = _____ g

    c. _____ g = 51.7 kg

10. Answer the following:

    a. Water boils at _____ °C.

    b. Would it be a nice day for the beach if it were 40°C? Explain.

Here are the answers:
Part B
1. a. 1.6
   b. 7.5
2. a. 2.3; 0.23
   b. 148.0; 14.80
   c. 87
   d. 1.250
   e. 7280
3. a. 4.6; 0.46; 0.046
   b. 2023.0; 202.30; 20.230
   c. 4217
   d. 3
4. a. 100
   b. 1000
5. 1.6 dm
6. a. 0.72
   b. 0.95
   c. 0.46; 0.0046; 0.000 046
7. Look at your model or the classroom model.
8. a. 40
   b. 4.2
   c. 82 800
9. a. 4.250
   b. 210
   c. 51 700
10. a. 100
    b. No, heat wave conditions.

Appendix I contains another form of the Metric Self-Test.

chapter
eight

# A CONTINUUM OF
# METRIC IDEAS

Like our customary system of measure, metric concepts, skills, and generalizations represent a system for recording measurements. Therefore, metric ideas can be directly substituted for customary concepts in the teaching program.

However, because the metric system demands a knowledge of decimals, decimal computation will need to be introduced earlier and stressed more heavily than it is now. A possible change in the mathematics program might be to place less stress on the common fractions, especially the algorithms, in order to provide the necessary time for teaching the decimal fractions.

A possible continuum of metric ideas might be as follows:

## A CONTINUUM OF METRIC CONCEPTS, SKILLS, AND GENERALIZATIONS

I. Kindergarten to Grade 3
   A. Mass
      1. Given a two-pan balance and two objects of unequal mass first guess which object is heavier, then verify your guess by the use of the balance.
      2. Given at least three objects of unequal size and mass, order them from lightest to heaviest, then verify the ordering by using a two-pan balance.
      3. Given a two-pan balance and some objects, identify those objects that balance each other.
      4. Given a two-pan balance and an object, find its mass to the closest 10 grams and then to the closest gram.
      5. Given a two-pan balance and sufficient weights,
         a. determine the number of grams that will balance 10 grams,

      b. determine the number of 10 grams that will balance a hectogram,

      c. determine the number of hectograms that will balance the kilogram.

6. Given a mass measure in whole kilograms, write the measurement in grams.

      example: 9 kilograms = _____ grams

B. Temperature

1. Given a thermometer, take and record to the closest 10 degrees the temperature of different things. examples:

    (a) the temperature in the classroom

    (b) the temperature of tap water

    (c) the temperature of your elbow

    (d) the temperature in the sun (outside)

2. Given a reading (Celsius scale) of a thermometer of an actual situation, state whether it is hot or cold.

3. Given a reading (Celsius scale) of a thermometer and a drawing of a thermometer, indicate the temperature on the drawing.

      example: Color to show the given temperature $30°C$.

C. Linear Measure

1. Given a metric ruler or metric tape measure and an object, write the measurement of its length to the closest decimeter; to the closest centimeter; to the closest meter.

2. Given a metric ruler or metric tape measure and an object, write the measurement of length of the object in decimeters and centimeters.

      example: Use your ruler to measure the length of your notebook.

              _____ decimeter and _____ centimeter

3. Given an object longer than a meter, write the measurement of its length to the closest meter.

4. Given the terms **centimeter, decimeter, meter,** in mixed order, arrange the units from shortest to longest.

5. Given a situation involving linear measure, determine the most appropriate unit of linear measure that should be used: centimeter, decimeter, or meter.

6. Given a metric ruler and an appropriate physical object,

find its perimeter to the closest centimeter; to the closest decimeter.
7. Given a metric ruler and two unequal line segments, find to the closest centimeter the difference between them.
8. Given the dimensions, in centimeters, of geometric figures, draw them.
    example:  A square with sides 3 cm long.
D. Area
1. Given a transparent grid marked into 1-centimeter squares and a rectangular surface, find the area of the surface.
2. Given a square centimeter and a rectangular region, measure the area of the rectangular region in square centimeters.
3. Given a rectangular region and a metric ruler, divide the region into square centimeters and determine its area.
4. Given the dimension in centimeter units of a rectangular region, draw and shade it.
5. Given the necessary supplies, draw or make a square meter.
E. Volume and Capacity
1. Given several 100-ml and 50-ml containers filled with the same substance and an empty liter container, determine the least number of 100-ml and 50-ml containers that will fill 0.250-liter, 0.500-liter, 0.750-liter, and 1-liter containers.
2. Given cubic centimeters and a cubic decimeter, determine the volume of the cubic decimeter by filling it with the cubic centimeters.
3. Given a rectangular prism, estimate its volume in cubic centimeters.  Then check by filling the container with cubic centimeters.
4. Given several common containers, for example, a soda pop bottle, find the capacity of each.
II. Grades 4 to 6
A. Mass
1. Given several common objects, for example, an orange, a pen, and a ball, estimate and then find the mass of each.
2. Given a liter of water, find its mass.
3. Given a list of metric mass units, arrange them in order from smallest to largest.  If possible, verify the ordering using weights.

4. Given a mass measure in grams, write an equivalent measure in kilograms and vice versa.
   example: 20 grams = 0.020 kilogram
5. Given an incomplete table of equivalent masses, supply the equivalent masses.
   example:
   Complete the following:

| | | | |
|---|---|---|---|
| Number of milligrams | 10 | 100 | _____ |
| Number of centigrams | _____ | 10 | 100 |
| Number of decigrams | 0.1 | _____ | 10 |
| Number of grams | _____ | 0.1 | 1 |
| Number of dekagrams | _____ | _____ | 0.1 |
| Number of hectograms | _____ | _____ | 0.01 |
| Number of kilograms | _____ | _____ | 0.001 |

6. Solve a word problem involving mass.
   example:
   Kimo wants to send six items in one package by airmail to his pen pal on the mainland. The items have masses of 737 g, 1280 g, 87 g, 2000 g, 746 g, 2015 g. The mailing rate for these is $1.79 per 1000 grams or fraction thereof.
   (a) How many grams does Kimo want to mail?
   (b) How much will it cost him?

B. Temperature
   1. Given a thermometer, record to the closest whole degree (+, −, or 0), the temperatures of different things.
   2. Given a drawing of a thermometer with an indicated temperature (+, −, or 0), record the temperature to the nearest whole degree.
   3. Given a reading of a thermometer (+, −, or 0), and a drawing of a thermometer, show the temperature on the drawing.
   4. Given a thermometer, record to the closest *tenth* of a degree (Celsius scale) the temperatures of different things.
   5. Given a drawing of a thermometer with an indicated temperature, record the temperature to the nearest tenth of a degree (Celsius scale).

6. Given the reference points of freezing, normal room temperature, boiling point, and body temperature, state a reasonable temperature in the Celsius scale.
7. Solve a real-life problem involving temperature.
C. Linear Measure
   1. Given a ruler marked with millimeters and an object to be measured, determine its measurement to the closest millimeter.
   2. Name two places that are approximately a kilometer apart.
      example: Name a place that is one kilometer from the school building.
   3. Estimate the distance from your home to school in kilometers.
   4. Given a stiuation involving linear measure, determine the most appropriate unit of linear measure: centimeter, millimeter, decimeter, meter, or kilometer.
   5. Measure a line segment to the closest millimeter, centimeter or decimeter. Then state what part it is of a meter.
      example: Measure the segment to the closest centimeter:

      _____

      What part of a meter is it?
   6. State equivalents, in other metric units of measure, of a kilometer.
      example: 9 kilometers = 9000 meters = 90 000 decimeters = 900 000 centimeters
   7. Solve a real life situation that involves computation with linear measures.
   8. Measure the length and width of a rectangle to the closest millimeter. Then find the perimeter of the rectangle.
   9. Given the dimensions of geometric figures in terms of centimeters or millimeters, draw them.
D. Area
   1. Given a square decimeter and square meter, determine by measuring how many square decimeters make a square meter.
   2. Given a square decimeter and a square centimeter, determine by measuring how many square centimeters make a square decimeter.

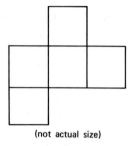

(not actual size)

**Figure 8.1** A region.

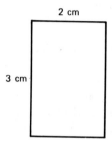

**Figure 8.2** A rectangular region.

3. Given a square centimeter, a square decimeter, and a square meter, determine by measuring how many square centimeters make a square meter.

4. Given a list of metric area units, arrange them in order from the smallest to the largest.

5. Given a square decimeter and a region, measure in square decimeters, the area of the region. An example is shown in Figure 8.1.

6. Given a square decimeter, a square centimeter, and a rectangular region, (a) measure in square centimeters and in square decimeters, the area of the region; (b) write the results as an equation.

   example: $2 \text{ dm}^2 = \underline{\hspace{1cm}} \text{ cm}^2$

7. Given a metric area measure in square centimeters, square decimeters, or square meters, find an equivalent metric area measure.

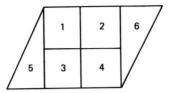

**Figure 8.3** Verifying the number of square centimeters.

example: Fill in the blank.

$$100 \text{ cm}^2 = \underline{\hspace{1cm}} \text{ dm}^2$$

8. Given the measures in cm, dm, or m units of the length and width of a rectangle, find its area. For example, students could be asked to find the area of Figure 8.2.
9. Measure in centimeters the base and altitude of a parallelogram, and find its area. Then verify by counting the number of square centimeters in the parallelogram. An example is shown in Figure 8.3.
10. Measure in centimeters the base and altitude of a triangle, and find its area. Then verify by counting the number of square centimeters in the triangle.
11. Measure in centimeters the radius of a circle, and find the area. Then verify by counting the number of square centimeters in the circle.

NOTE: Other area objectives may be added. In all cases, instead of customary linear measures (inches, feet, etc.) we use metric linear measures (cm, dm, m, etc.).

E. Volume and Capacity
   1. Given several 10-mℓ containers filled with the same substance and an empty liter container, determine the least number of 10 mℓ containers that will fill 0.1 ℓ, 0.2 ℓ, 0.3 ℓ, ..., 0.9 ℓ, 1.0 ℓ. Next determine how many milliliters would be needed to fill 0.1 ℓ, 0.2 ℓ, 0.3 ℓ, ..., 0.9 ℓ, 1.0 ℓ.
   2. Given capacity containers (deciliter, centiliter, liter, and dekaliter) and a source of water, sand, or rice, determine what part of a liter a centiliter and a deciliter comprise. Next determine what part of a dekaliter a liter comprises.

3. Given the capacity units (milliliter, deciliter, centiliter, dekaliter, and hectoliter), order them from smallest to largest.

4. Express a given capacity unit (milliliter, centiliter, deciliter, or hectoliter), in terms of a liter.
   example:
   Fill in the blanks.

   1 deciliter   = _____ liter

   0.2 milliliters = _____ liter

5. Given a capacity measure in liters, state its equivalence in milliliters, and given a capacity measure in milliliters, state its equivalent in liters.
   example:
   2 liters        = 2000 milliliters
   200 milliliters = 0.200 liter
   0.020 liter     = 20 milliliters

6. Given capacity measures of less than a 1000 milliliters expressed in liters (that is, in decimal form), arrange them in order from least to most.
   example:  Arrange these measures from least to most.
   0.001 ℓ,                0.010 ℓ,                0.100 ℓ

7. Given a cubic decimeter, measure its height, width, and depth. Next determine its volume by filling it with cubic centimeters. State a relationship between its volume and its height, width, and depth.

8. Given a rectangular prism, measure its height, width, and depth. Next determine its volume by filling it with cubic centimeters. State a relationship between the prism's volume and its height, width, and depth.

9. Given the dimensions (in metric measures) of a rectangular solid, find its volume.

10. Given a cubic decimeter, filled with water, determine the capacity of 100, 200, 300, . . . , 900, and 1000 cubic centimeters of water by filling a graduated liter container.

11. Given the volume measures, express their equivalents in capacity measures and vice versa.
    example:
    Fill in the blanks.

(a)  $1 \text{ cm}^3$ = _____ mℓ

(b)  $20 \text{ cm}^3$ = _____ mℓ

(c)  $500 \text{ cm}^3$ = _____ ℓ

(d)  0.5 liter = _____ $\text{cm}^3$

12. Given the necessary supplies, construct a cubic meter. Then estimate the number of cubic decimeters it will contain.

## Exercises

1. Review and report on the metric scope and sequence of any elementary textbook series.

2. Obtain curriculum guides from several school systems and compare the scope and sequence of their metric programs.

3. Evaluate the continuum provided in the text. Do you agree with the grade placement of the objectives?

4. Determine and share with the class how you might evaluate whether the objectives of the continuum are realized.

5. Review a film, filmstrip, or tape cassette and discuss how, if at all, it would enhance the attainment of the objectives of the metric continuum.

6. Play a metric game. See Appendix II. Where in the continuum does it belong?

7. Make a bulletin board display designed to enhance some objective(s) of the metric continuum.

8. Discuss the advantages and disadvantages of learning two systems of measurement, customary and metric, at the same time.

9. Given the metric continuum of this chapter, where would you place decimal fraction concepts?

# chapter
# nine

# SETTING UP A CLASSROOM LABORATORY FOR TEACHING METRIC UNITS

## CLASSROOM MANAGEMENT

In this book we have suggested a laboratory approach to teaching the metric system. With this approach students first perform simple experiments using metric measures, slowly building their own concepts regarding these units of measure. Once students are familiar with the units of measure, they use their new knowledge by using it to solve problems that are presented in a concrete manner. An example of such a problem is finding the area or perimeter of the classroom once students have become familiar with linear metric units. In addition to physically performing the measuring act, students record their data and compare their findings with the findings of other students. This approach allows students to be active participants, in contrast to a lecture-textbook approach in which they are passive spectators. The laboratory approach, because it takes students from the simple concrete situation to the more formal symbolic problem solving, permits students to work at their own pace and set their own sequence for learning. The concrete nature of the materials allows the materials and activities to do the teaching. The teacher assists, observes, and brings groups of students into applying this knowledge to solve computation and word problems using only the mathematical symbols presented through worksheets and the textbooks.

Many teachers teach their classes as groups or possibly subdivide the class into two or three ability groups. The question that arises for most teachers is: "How do I move from the formal textbook teaching to a laboratory approach?" To make this switch without training the students can be chaotic. Therefore, we would like to suggest the following transitional phase as a possible model for bridging the change in teaching strategy. We suggest that you instruct a small group of students in using the laboratory techniques, while the remainder of the students work on their regular materials in mathematics. Here are some activities for laboratory instruction that you might use in getting started.

### Linear Metric Activities

### Nonstandard Units

1. Measure the length of your desk using
   a. any crayon
   b. the width of your hand
   c. any pencil
   Which measurement is correct? Which measure is best? Compare your measurements with your neighbors. Name other things you can use to measure the length of your desk.
2. Name things you could use to measure the distance around the school building.
3. How would you measure the length of a tall building?

### Standard Units

1. Measure your book using
   a. the paper clip
   b. the popsicle stick
   c. the penny
   Compare your measurements with your neighbors. Which measure is correct? Which unit of measure is the best? Name other objects that could be used to measure the book. Are they as good or better than those given?
2. Name standard units you would use to measure
   a. the height of the water fountain
   b. "pick-up" sticks
   c. a roll of Life Savers candy
   d. shoelace

3. Which standard unit is best for measuring your favorite toy?
   a. the popsicle stick
   b. the classroom eraser
   c. a regular postage stamp
   d. none of the above
   If you chose "none of the above," name your own standard.

## Meter

1. Using a blank (without markings) meter stick measure:
   a. how far a paper airplane flies
   b. how far you go on the slide
   c. how far you can throw a ball
   d. how far you can jump
   e. how far away is the principal's office

2. Estimate, then measure three long or tall things outdoors.

| Items | Estimate | Measurement |
|---|---|---|
| a. | _____ m | _____ m |
| b. | _____ m | _____ m |
| c. | _____ m | _____ m |

3. With a partner name three things. Then each of you estimate the lengths of the items. Now each measures to see who is the better estimator.

| Item | My Estimate | Partner's Estimate | Measurement |
|---|---|---|---|
| a. | _____ m | _____ m | _____ m |
| b. | _____ m | _____ m | _____ m |
| c. | _____ m | _____ m | _____ m |

4. Name three things that are slightly longer than a meter. Name three things that are slightly shorter than a meter. Check by measuring.

## Decimeter

1. Cut a piece of tagboard or adding machine tape that is one decimeter in length. (See Figure 9.1.)
   a. How many decimeters long is a meter?

Figure 9.1  One decimeter.

b.  1 m = _____ dm
c.  What part of the meter is the decimeter?

2.  Using your decimeter, measure three things of interest to you.
Have a friend check your measurements.

| Item | My Measurement | Friend's Measurement |
|---|---|---|
| a. | _____ dm | _____ dm |
| b. | _____ dm | _____ dm |
| c. | _____ dm | _____ dm |

3.  Estimate, then measure the following:

| Item | Estimate | Measurement |
|---|---|---|
| a. height of a large stuffed animal, doll, or puppet | _____ dm | _____ dm |
| b. handle of a wagon | _____ dm | _____ dm |
| c. distance around a ball | _____ dm | _____ dm |

4.  Name three things that are slightly longer than a decimeter.
Name three things that are slightly shorter than a decimeter.  Check
by measuring.

## Centimeter

1.  Divide your decimeter into 10 equal parts.  (See Figure 9.2.)
Each part is one centimeter long.  How many centimeters make a
decimeter?

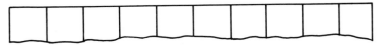

Figure 9.2  A decimeter divided into 10 equal parts.

2.  Use your newly marked decimeter to measure the following
to the closest centimeter.
a.  height of a jar of paste or glue

    b. width of your thumb
    c. length of a toy car, truck, or bus
    d. width of a playing record
    e. length of a paint brush

3. Estimate the length of three things at home to the closest centimeter. Check by measuring.

| Item | Estimate | Measurement |
|------|----------|-------------|
| a. | _____ cm | _____ cm |
| b. | _____ cm | _____ cm |
| c. | _____ cm | _____ cm |

## Meter, Decimeter, Centimeter

1. Make a meter using adding machine tape, heavy paper, ribbon, or cord.
    a. Draw in the decimeter marks.
    b. Draw in the centimeter marks.
    c. How many decimeters make a meter? How many centimeters make a decimeter? How many centimeters make a meter?

    d. _____ cm = _____ dm = _____ m
2. Estimate, then measure
    a. how high you can swing
    b. how high you can climb
    c. how high you can reach
3. Estimate, then measure
    a. the height of members of your family
    b. the height of your pet
    c. the height of your bed

*Diagram I*

**How to Start: First Day**

| Regular or review work in mathematics that requires minimal or no teacher supervision | Laboratory instruction using linear measurement |
|---|---|
| BBBBB    CCCCC<br>DDDDD    EEEEE | AAAAA |

Step 1. Teacher assigns work to groups B, C, D, E.
Step 2. Teacher takes group A of five or more students to the laboratory area, which is within the classroom, and introduces them to linear measurement using concrete materials.
Step 3. Once group A is working satisfactorily the teacher returns to the other groups supervising their regular work.
Step 4. Near the end of the mathematics period the teacher returns to the laboratory area and evaluates with group A.

*Diagram II*

**How to Start: Second Day**

| Regular or review work | Laboratory Area |
|---|---|
| in mathematics that | AB, AB, AB, AB, AB |
| requires minimal or no | |
| teacher supervision | |
|     CCCCC | |
|     DDDDD | |
|     EEEEE | |

Step 1. Teacher assigns regular or review work to groups C, D, E.
Step 2. Teacher matches children in groups A and B, pairing those who worked with the metric materials the previous day with those who have not worked with the materials.
Step 3. Group A children help group B children explore with metric material. Teacher returns to groups C, D, E.
Step 4. Near the end of the math period the teacher evaluates with partner groups in the laboratory area.

*Diagram III*

**How to Start: Third Day**

| Regular or review work | Laboratory Area AAAAA |
|---|---|
| in mathematics that | works on linear application |
| requires minimal or no | BC, BC, BC, BC, BC form |
| teacher supervision | partner groups |
|     DDDDD | |
|     EEEEE | |

Step 1. Teacher assigns regular or review work to groups D and E, and releases group A to work on applications of linear measurement.

Step 2. Children in groups B and C form partners, pairing those who worked with the metric materials the previous day with those who have not worked with the materials.

Step 3. Group B children help group C partners, teacher supervises in laboratory area.

Step 4. Teacher returns to groups D and E and works with them.

Step 5. Near the end of the period the teacher returns to the laboratory area and evaluates.

*Diagram IV*

**How to Start: Fourth Day**

> Laboratory Area BBBBB works on linear application
> CD, CD, CD, CD, CD partner groups
> AE, AE, AE, AE, AE partner groups

Step 1. Teacher sends group B to laboratory area to work on linear application.

Step 2. Children in groups C and D form partner groups, teacher releases to laboratory area.

Step 3. Children in groups A and E form partner groups, teacher releases and supervises all groups.

Step 4. Near the end of mathematics period the teacher plans with each group the possible sequence they may wish to pursue during the succeeding weeks.

The possible sequence for the groups is shown on the following page:

The preceding diagram shows that each group's sequence is somewhat different, so a minimal amount of special equipment is in use at any time. It should be recognized that some grades, especially K-2, and for some groups of students, the groups would not reach the formal symbolic applications and possibly not even the concrete applications. Primary teachers may need to work directly with groups for longer periods of time. Most of the material presented in this manual requires that students be able to read. For those children who cannot read, the teacher may wish to make audio tapes, or match these children with children who can read. The teaching diagrams outlined here represent one of many ways to start a laboratory

| Group A | Group B | Group C | Group D | Group E |
|---|---|---|---|---|
| meter | meter | meter | meter | meter |
| decimeter | decimeter | decimeter | decimeter | decimeter |
| centimeter | centimeter | centimeter | centimeter | centimeter |
| millimeter | millimeter | millimeter | millimeter | millimeter |
| applications of linear measure | temperature | applications of linear measure | applications of linear measure | temperature |
| formal problems | applications of linear measure | capacity | mass | applications of linear measure |
| capacity | mass | temperature | temperature | mass |
| mass | capacity | mass | capacity | capacity |
| temperature | formal problems | formal problems | formal problems | formal problems |

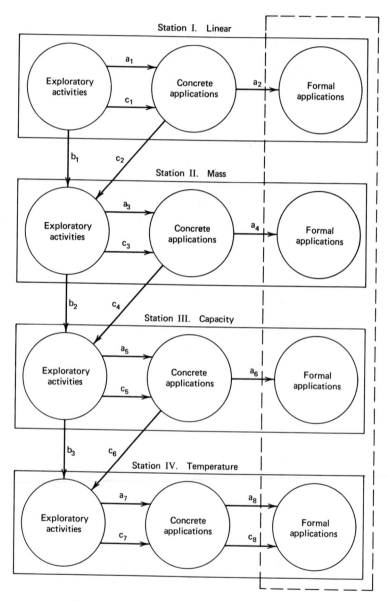

**Figure 9.3** Traffic patterns among stations.

approach. The amount of structure and teacher guidance will depend on the amount of previous independence the students have had.

A decision also needs to be made regarding working space. If the students can work directly at the learning station this makes participation and storage of materials very easy. However, in an open classroom where there are multiple learning stations and space is limited, the students may need to take materials from the learning station to a work area. Traffic patterns and work areas need to be established, otherwise the mere mechanics of trying to get 15 or 30 students to work independently becomes overburdening.

In setting up the physical arrangement of a laboratory area, the teacher should cluster the learning stations so that, when appropriate, the physical layout leads the students through each of the three stages of development: experiment, concrete application, and formal application. Figure 9.3 shows how a teacher might set up the stations in her classroom. The arrows indicate the posible flow of activities: "a's" represent one possibility, "b's" another, and "c's" a third possibility. Possibility "b" is more likely to be for K-2 students who have difficulties with concrete application.

## CONTENT STRATEGY

The previous section discussed the classroom management aspects of using a laboratory approach. We now suggest two possible ways for introducing content.

### STRATEGY 1

The open-ended strategy in which the students recognize the need for standardized units.

The teacher sets up a series of measurement activities for length, mass, and capacity. The open-ended questions take the following pattern: Use a pencil, your hand, a book, or any other item, and measure how long or high the following are: your friend, the height of your desk, and so forth. A similar set of questions are made for mass and capacity. The students measure and record their findings. A discussion is held and a chart is made combining the findings of the class.

The teacher then gives them a second set of questions. In this set, the students remeasure all of the same items as in the first task, but are given a common unit of measure. For example, in measuring length the teacher gives each small group of students a new, unsharpened pencil as their unit of measure. For mass, it would be paper clips, and for capacity, the box that held the paper clips.

The teacher then compares findings using a standard unit with the first set of measurements. The students may be led to understand the need for a standard unit. This set of activities then leads into the unit on metric measures.

### STRATEGY 2

This strategy starts with students becoming familiar with the metric units. The beginning point would be the linear measurement unit. The teacher sets up all the learning stations with material for linear, mass, capacity, and temperature measurements and then begins with the class as a whole. The students start with the meter and continue on their own through the linear unit. Once they complete the linear unit (at least the exploratory stage) they may then work on any of the other three units (mass, capacity, temperature).

In order to assure learning and provide the teacher with an opportunity for informal assessment, the teacher and small groups of students should continually discuss the activities the students are working at.

### EVALUATION

The laboratory approach encourages individualization, providing opportunities for students to select activities and to work at their own pace. This type of learning creates assessment problems for the teacher. Therefore, an evaluation sheet should be used by the student, serving as a guide and providing a focus for a student-teacher evaluation conference. Following is a possible evaluation sheet.

### SAMPLE STUDENT EVALUATION SHEET

Name:                                                          Check
Meter                                                          Off
    Task 1   Show someone how long a meter is.

Task 2  State how many foot lengths a meter comprises.
Task 3  Demonstrate a meter stride.
Task 4  Estimate the length of three objects to the nearest meter.

Decimeter
Task 1  State how many decimeters equal a meter.
Task 2  Estimate the length of three objects to the nearest decimeter.
Task 3  Estimate the length of three objects in terms of meters and decimeters.
Task 4  Define "deci."

Centimeter
Task 1  State how many centimeters long my thumb is.
Task 2  Estimate the length of three objects to the nearest centimeter.
Task 3  State the length of sides, *a, b, c,* in centimeters.  (See Figure 9.4.)

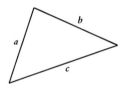

**Figure 9.4**  Triangle.

Task 4  Define "centi."
Task 5  State how many centimeters equal a decimeter.
Task 6  State how many centimeters equal a meter.
Task 7  Rewrite the following in decimal form in terms of meters.
  (a) 2 m and 85 cm
  (b) 3 m and 5 cm
  (c) 96 cm

Kilometer
Task 1  Name a place that is a kilometer away from here.

Perimeter
Task 1  Measure and record the perimeter of a desk.
Task 2  Measure and record the perimeter of the room.
Task 3  Measure and record the perimeter of the bulletin board.

Area
  Task 1   Draw a square decimeter.
  Task 2   Draw a square meter.
  Task 3   Draw a square centimeter.
Volume
  Task 1   Construct a cubic centimeter.
  Task 2   Construct a cubic decimeter.
  Task 3   Name the number of cubic centimeters in a cubic decimeter.
Temperature
  Task 1   Determine my body temperature.
  Task 2   Determine hot tap water temperature.
  Task 3   Determine cold tap water temperature.
Mass
  Task 1   Estimate and measure the mass of three common objects.
Capacity
  Task 1   Estimate and measure the capacity of three common objects.

### EXERCISES

1. Assume you have $50 to spend on metric supplies and equipment. What would you purchase and why?

2. List things fourth graders would find interesting to measure.

3. Read and report to the class: *The Laboratory Approach to Mathematics* by Kidd, Myers, and Cilley.

4. Read and report to the class: "Arranging a Metric Center in Your Classroom" by Bruni and Silverman in *Arithmetic Teacher,* January 1976.

5. Play one of the games listed in Appendix II and share with the class its possible use in a laboratory setting.

6. Write up task cards that provide exploratory activities with mass for primary grades.

7. Locate and review some metric task cards. See Appendix II.

8. Make up a sample evaluation sheet on metric competencies that could be used in the primary grades.

9. Review several elementary textbook series and note when treating temperature whether they used Strategy 1 or Strategy 2 as discussed in the text.

10. Characterize the "laboratory approach" to teaching.
11. Develop one or more metric learning stations for an elementary classroom.
    a. Write several task cards to be used at the stations.
    b. Obtain and/or make the supplies and equipment to be placed at the stations. For example, sample square decimeters, gram weights, and thermometers.
    c. Determine the possible traffic flow in the classroom and at the stations. Sketch a floor plan. Write down what each group of students will be doing at particular periods of time, and indicate on the floor plan the anticipated whereabouts of the several activities.

# appendix
## i

# METRIC SELF-TEST

**PART A**

1. Estimate:
   a. The width of this page
   b. The area of a large index card
   c. The volume of a shoe box
   d. The capacity of a 7-Up can
   e. The mass of your wallet or coin purse
2. Name objects in your immediate environment with the following approximate measurements.
   a. 15 centimeters
   b. 10 square centimeters
   c. 0.25 liter
   d. 1 kilogram
   e. 10 cubic centimeters
3. Check your answers to 1 and 2 above by actually measuring.

**PART B**

1. Write these measurements in abbreviated form using the meter as the unit of length.

   a. 89 decimeters = _____ m

   b. 3 meters and 4 decimeters = _____ m
2. Fill in the blanks.

   a. 9 cm = _____ dm = _____ m

   b. 857 cm = _____ dm = _____ m

   c. 0.7 dm = _____ m

   d. 6.4 dm = _____ cm

   e. 72.8 m = _____ cm

3. Fill in the blanks.

    a. 685 mm = _____ cm = _____ dm = _____ m

    b. 1023 mm = _____ cm = _____ dm = _____ m

    c. 0.031 m = _____ mm

    d. 0.8 m = _____ mm

4. Answer the following:

    a. A dekameter is equivalent to _____ meters.

    b. A hectometer is equivalent to _____ meters.

5. Determine the perimeter of a triangle with sides 0.004 m, 0.005 m, and 0.006 m.

6. Complete the following statements:

    a. 10 $mm^2$ = _____ $cm^2$

    b. 56 $cm^2$ = _____ $m^2$

    c. 24 $mm^2$ = _____ $cm^2$ = _____ $dm^2$ = _____ $m^2$

7. Draw a picture of a cube that has a volume measure of one cubic decimeter.

8. Complete the following statements:

    a. 1 mℓ = _____ ℓ

    b. _____ mℓ = 0.20 ℓ

    c. 0.570 ℓ = _____ mℓ

9. Fill in the blanks:

    a. 300 g = _____ kg

    b. _____ g = 6.20 kg

    c. 0.08 kg = _____ g

10. Answer the following:

    a. Water freezes at _____ °C.
    b. Would you wear a coat if it were 97°C? Explain.

Here are the answers
**PART B**
1. a. 8.9
   b. 3.4

2. a. 0.9; 0.09
   b. 85.7; 8.57
   c. 0.07
   d. 64
   e. 7280
3. a. 68.5; 6.85; 0.685
   b. 102.3; 10.23; 1.023
   c. 31
   d. 800
4. a. 10
   b. 100
5. 0.015
6. a. 0.10
   b. 0.0056
   c. 0.24; 0.0024; 0.000 024
7. Look at your model or the classroom model.
8. a. 0.001
   b. 200
   c. 570
9. a. 0.3
   b. 6200
   c. 80
10. a. 0
    b. No, too warm.

# appendix
## ii

# INSTRUCTIONAL
# AIDS AND IDEAS

## A. SUPPLIES AND EQUIPMENT

On the next few pages are listed some commercial supplies and equipment, along with approximate prices and vendors, that are helpful in getting started in teaching the metric system. This is only a partial listing and should not be construed as an endorsement of specific vendors.

Addresses of the vendors are:

Creative Publications, Inc., P. O. Box 10328, Palo Alto, CA 94303

Dick Blick, P. O. Box 1267, Galisburg, IL 61401

Ideal School Supply Co., 11000 S. LaVerne Ave., Oak Lawn, IL 60453

La Pine Scientific Co., 920 Parker St., Berkeley, CA 94710

Sargent-Welch Scientific Co., 1617 East Ball Rd., Anaheim, CA 92803

Sigma Scientific, Inc., P. O. Box 1302, Gainsville, FL 32601

Table Appendix IIA  Supplies and Equipment
(A Starter Set of Classroom Materials)

| Object | Approximate Cost | Vendor and Catalog Number | Description |
|---|---|---|---|
| *For linear measurement:* | | | |
| Meter stick | $ 2.90 | Dick Blick 88007 | Wooden, only cm marks (numbered) |
| Meter stick | 3.60 | Dick Blick 88021 | Wooden, cm and mm (numbered) marks |

Table Appendix IIA continued

| Object | Approximate Cost | Vendor and Catalog Number | Description |
|---|---|---|---|
| Metric ruler | $ 1.95 (36 rulers) | Creative Publications | Heavy cardboard, 25 cm long; one edge only cm marks (numbered); other edge cm and mm marks |
| Tape measure | 2.60 (10 tapes) | Dick Blick 88010 | Coated linen with metal ends 100 cm long; cm and mm marks; cm numbered |
| Metric ruler | 3.75 (20 rulers) | Creative Publications 36415 | Flexible vinyl 30 cm long; one edge cm and dm marks; other edge cm and mm marks |
| Tape measure | 3.51 (100 strips) | La Pine z-12358 | Paper, 1 meter long; marked in cm on one side |
| *For area measurement:* | | | |
| Transparent grid | 4.25 (10) | Creative Publications 36500 | Acetate; cm grids 25 cm X 25 cm |
| Transparent grid | 10.80 (10) | Dick Blick 88171 | Clear plastic; mm grids; 250 mm X 250 mm |
| *For volume measurement:* | | | |
| Centimeter base ten | 19.90 (a set) | Dick Blick 88152 | Plastic colored blocks; set includes a cubic dm, 10 flats (square dm), 10 sticks (10 cm long), 100 cubes (cm) |
| Cubic liter | 2.60 | Dick Blick 88134 | Clear; graduated in 100 ml divisions |

Table Appendix IIA continued

| Object | Approximate Cost | Vendor and Catalog Number | Description |
|---|---|---|---|
| *For capacity measurement:* | | | |
| Beakers | $ 8.50 (a set) | Creative Publications 36652 | Translucent; has pouring spout and is clearly calibrated; set includes 1 each of 10 mℓ, 100 mℓ, 250 mℓ, 600 mℓ, and 1 liter |
| *For mass measurement:* | | | |
| Balance | 28.50 | Ideal School Supply | Ohaus brand; includes weights |
| Balance | 23.75 | Creative Publications 36710 | Ohaus brand, does not include weights |
| Weights | 22.25 (a set) | Sargent-Welch MS-4285-D | Ohaus brand, precision brass weights; 12 weights: 1 to 500 grams |
| Bath scale | 10.00 | Creative Publications 36726 | Marked in kg only |
| *For temperature measurement:* | | | |
| Thermometer | 1.20 (2 per set) | Sigma 300102 | Celsius and Fahrenheit scale; $10^{\circ}$C divisions from $-20^{\circ}$C to $110^{\circ}$C; metal backing |

## B.  RESOURCE MATERIALS

Some sources of recently produced materials that will aid you in integrating metric into your curriculum are listed on the following page.

## Charts

Charts serve as a constant reinforcer of classroom activities. They are available in poster form. The following sources carry posters that are especially appropriate for the elementary school classroom. They range in price from $1.50 to $10.00.

Activity Resources Co., Inc., P. O. Box 4875, Hayward, CA 94540

A/V Instruction Systems, P. O. Box 191, Somers, CT 06071

Central Instrument Co., 900 Riverside Dr., New York, NY 10032

Creative Publications, Inc., P. O. Box 10328, Palo Alto, CA 94303

Educational Teaching Aids, 159 W. Kinzie St., Chicago, IL 60610

Hayes Co., 321 Pennwood Ave., Wilkensburg, PA 15221

Ideal School Supply Co., 11000 S. LaVerne Ave., Oak Lawn, IL 60453

La Pine Scientific Co., 920 Parker St., Berkeley, CA 94710

Math Group, Inc., 396 E. 79th St., Minneapolis, MN 55420

Math-Master, P. O. Box 1911, Big Spring, TX 79720

Milton Bradley Co., Springfield, MA 01101

Modern Mathematics Materials, 1658 Albemarle Way, Burlingame, CA 94010

National Micrographics Association, 8728 Colesville Rd., Silver Spring, MD 20910

Western Learning Laboratories, 11923 Venice Blvd., Los Angeles, CA 90066

## Kits

Kits generally consist of a teacher's guide, a student workbook, and the equipment and supplies needed to carry out the activities suggested in the student workbook. It is a time saver for those who cannot search out individual pieces of supplies and equipment. Kits range in price from about $25 to $450. The higher-prices kits generally contain additional audiovisual materials. For each company the approximate prices of kits are given.

Addison-Wesley Publishing Co., Inc., Sand Hill Rd., Menlo Park, CA 94025 ($70)

AERO Educational Products, P. O. Box 71, St. Charles, IL 60172 ($50)

Dick Blick Co., P. O. Box 1267, Galisburg, IL 61401 ($150)

Coronet, 65E S. Water St., Chicago, IL 60601 ($200)

Creative Publications, Inc., P. O. Box 10328, Palo Alto, CA 94303 ($25)

Creative Teaching Associates, P. O. Box 7714, Fresno, CA 93727 ($145)

Educational Teaching Aids Division, 159 W. Kinzie St., Chicago, IL 60610 ($64)

Enrich, 3437 Alma St., Palo Alto, CA 94306 ($110)

Imperial International Corp., P. O. Box 548, Kankakee, IL 60901 ($450)

La Pine Scientific Co., 920 Parker St., Berkeley, CA 94710 ($200)

Leicestershire Learning Systems, Box MS, New Gloucester, ME 04260 ($82.50)

Listener Educational Enterprises, 6777 Hollywood Blvd., Hollywood CA 90028 ($325)

Math-Master, P. O. Box 1911, Big Spring, TX 79720 ($85)

Media Materials, Inc., 2936 Remington Ave., Baltimore, MD 21211 ($100)

Charles D. Merrill Publishing Co., 1300 Alum Creek Dr., Columbus, OH 43216 ($50)

Schloat Productions, A Prentice-Hall Company, 159 White Plains Road, Tarrytown, NY 10591 ($317)

Science Research Association, 259 E. Erie St., Chicago, IL 60611 ($85)

Silver Burdett Co., Morristown, NJ 07960 ($100)

Singer Education Division, Society for Visual Education, 1345 Dinersey Pkwy, Chicago, IL 60614 ($134)

Weber Costello, 1900 N. Narragansett Ave., Chicago, IL ($43)

**Task Cards**

To set up learning centers or stations, to provide individual or group activities, or to vary instructional media, task cards are very helpful. Many kits include task cards. The companies named below have task cards separate from kits. A packet of task cards generally costs less than $10.

Activity Resources Company, Inc., P. O. Box 4875, Hayward, CA 94540

BHU Company, 23358 Hartland St., Canoga Park, CA 91307

Creative Teaching Press, Inc., 1900 Tyler Ave., Suite 22, El Monte, CA 91733

Houghton Mifflin, 777 California Ave., Palo Alto, CA 94304

La Pine Scientific Company, 920 Parker St., Berkeley, CA 94710

Love Publishing Company, Denver CI 90222

Math Group, Inc., 396 E. 79th St., Minneapolis, MN 55420

Michigan Council of Teachers of Mathematics, 2165 E. Maple Road, Birmingham, MI 48008

Prentice-Hall Learning Systems, Inc., P. O. Box 47x, Englewood Cliffs, NJ 07632

Selective Educational Equipment, Inc., 3 Bridge St., Newton, MA 02195

Teachers, P. O. Box 398, Manhattan Beach, CA 90266

Thinkisthenics, P. O. Box 1180, Palm Springs, CA 92262

## Audiovisuals

Audiovisual aids include film loops, films, filmstrips, and tape cassettes. Generally included with these aids are a teacher's guide and a student booklet. The initial cost of audiovisuals is generally high. Films range in cost from about $90 to $200; filmstrips range from $30 to $165 per set. The following companies offer materials intended for the elementary school level. Many of the filmstrips and films have either a story line, cartoon, or the child's world approach. Many of them have been reviewed in the *Arithmetic Teacher.*

## Filmstrips

BFA Educational Media, 2211 Michigan Ave., Santa Monica, CA 90404

R. W. Bruce Co., 1401 Mount Royal Ave., Baltimore, MD 21217

Coronet Instructional Media, 65 E. South Water St., Chicago IL 60601

Creative Learning, Box 4-D, 19 Market St., Warren, RI 02885

Denoyer-Geppert, 5235 Ravenswood Ave., Chicago, IL 60640

Educational Activities, Inc., P. O. Box 392, Freeport, NY 11520

Education Projections Co., 3070 Lake Terrace, Glenview, IL 60025

Encyclopedia Britannica Educational Corporation, 425 N. Michigan Ave., Chicago, IL 60611

Eye Gate, 146-01 Archer Ave., Jamaica, NY 11435

Film Communication, P. O. Box 113, North Field, IL 60093

Films Incorporated, 1144 Wilmette Ave., Wilmette, IL 60091

Houghton Mifflin Co., One Beacon St., Boston, MA 02107

Imperial Educational Resources, P. O. Box 5500, 202 Lake Miriam Dr., Lakeland, FL 33803

La Pine Scientific Co., 920 Parker St., Berkeley, CA 94710

Library Filmstrip Ctr., 3033 Aloma, Wichita, KA 67211

Math-Master, P. O. Box 1911, Big Spring, TX 79720

Pathescope Educational Media, Inc., 71 Weyman Ave., New Rochelle, NY 10802

Society for Visual Education, Inc., 1345 Diversey Pkwy, Chicago, IL 60614

## Films

Barr Films, P. O. Box 7-C, Pasadena, CA 91104

Stephen Bosustow Productions, 1649 Eleventh St., Santa Monica, CA 90404

Coronet Instructional Media, 65 E. South Water St., Chicago, IL 60601

Davidson Films, Inc., 3701 Buchanan St., San Francisco, CA 94123

Film Communication, P. O. Box 113, North Field, IL 60093

La Pine Scientific Co., 920 Parker St., Berkeley, CA 94710

Oxford Films, 1136 N. Las Palmas Ave., Los Angeles, CA 90038

Perennial Education, Inc., 1825 Willow Rd., Northfield, IL 60093

## Converters

Edmunds Scientific, 380 Edscorp Bldg., Barrington, NJ 08007

Jaydee Specialties, P. O. Box 536 Wilmette Ave., Wilmette, IL 60091

Kilm Manufacturing Co., 3151 U. L. 33 North, Benton Harbor, MI 99022

Perrygraf, 2215 Colby Ave., Los Angeles, CA 90064

Pickett, 17621 Von Karman Ave., Irvine, CA 92705

Sterling Plastics, Sheffield St., Mountain Side, NJ 07092

Telex, 9600 Aldrich Avenue S., Minneapolis, MN 55420

Union Carbide Corp., Educational Aids Dept., P. O. Box 363-B, Tuxedo, NY 10987

Vari-vue International Inc., 650 South Columbus Ave., Mount Vernon, NY 10550

## Student Workbooks

To supplement or replace your present textbook's work on measurement, student workbooks can be very helpful. Teacher's guide generally accompany workbooks. The following sources have relatively complete measurement programs. Their booklets are suitable for the elementary school grades. All are activity-oriented and avoid comparison of metric units with English units.

Addison-Wesley Publishing Co., Inc., Sand Hill Rd., Menlo Park, CA 94025

Harcourt Brace Jovanovich, 757 Third Avenue, New York, NY 10017

Hawaii Metric Project, 1776 University Avenue, Honolulu, HI 96822

Laidlaw Brothers, River Forest, IL 60303

McGraw-Hill, Inc., 1221 Avenue of the Americas, New York, NY 10020

Random House, Inc., 201 E. 50th St., New York, NY 10022

Scott Foresman & Co., 1900 E. Lake Ave., Glenview, IL 60025

## Student Games

To reinforce learning, metric games provide needed practice in a form students will enjoy.

Creative Teaching Associates, P. O. Box 293, Fresno, CA 93708

*Merry Metric:* Reinforces the meanings of the prefixes and their relationship to meter, liter, and grams. $3.50

*Metric Derby:* Provides "hands on" experience in metric measure, drill and practice in multiplication, decimals, and metric prefixes. $5.95

*Metric Challenge:* Can be used for drill in decimals or metric measures. $4.94

*Metric Spinner Games:* Helps you learn about meters, liters, and grams; metric prefixes; and metric equivalents. $6.95

Educational Teaching Aids, 159 W. Kinzie St., Chicago, IL 60610

*Metric Bingo:* A game designed for the gradual learning of metric terms.

Education Plus, 18584 Carlwyn Dr., Castro Valley, CA 94546

*Metric Monster:* A card game that familiarizes students with metric units. $1.50

Great Ideas Inc., P. O. Box 274, Commack, NY 11725

*Metric-Tac-Toe:* A tic-tac-toe game with cubes marked with metric terms to develop an understanding of the metric system. $6.50

*Kilo:* A game played with cubes and similar in format to Bingo. $6.50

*Metrideck:* A deck of metric cards played in game formats similar to Rummy and other common card games. $4.95

W. J. Hampton Real-T-Facs, 26 Overlook Dr., P. O. Drawer 449-B, Warwick, NY 10990

*Meter-Liter-Gram:* The game board is laid out in metric units, the game tokens are cubic centimeters, and players move as they respond to metric questions. The game teaches metric measurement to children. $10.00

Ideal School Company, 11000 S. LaVerne Ave., Oak Lawn, IL 60453

*Metric Match:* A board game designed to reinforce understanding of the relationships of the various units of metric measurements, their prefixes, and symbols.

Kent Educational Services, P. O. Box 903, Oviedo, FL 32765

*Metric Ladder Race:* A board game that provides practice in using metric prefixes. Only linear units are used. $7.95

Lawhead Press Inc., 900 E. State Street, Athens, OH 45701 (also available from Damon/Educational Division, 80 Wilson Way, Westwood, MA 02090)

*Decimeter:* Players move their cubes around the board, stopping at various stores along the way making purchases. The game tends to strengthen knowledge of metric units and prefixes. $12.50

Math Group Inc. 396 E. 79th St., Minneapolis, MN 55420

*Metric "21":* A card game played like "21" or "Blackjack." Students estimate lengths. $2.50

*Match-a-Gram:* A card game wherein players match cards consisting of an object, its mass, and an equivalent expression of the mass. $2.95

*Match-a-Meter:* Similar to *Match-a-Gram.* $2.95

*Match-a-Liter:* Similar to *Match-a-Gram.* $2.95

Metrix, P. O. Box 19101, Orlando, FL 32814

*Metrix Match:* Players match (a) prefixes with decimal equivalents, (b) measurement types with basic metric units, (c) metric unit names with symbols. $5.99

*Metrication:* Similar to Monopoly and other buying and selling games. It helps players to memorize the prefixes of the system. $10.00

W L W Enterprises, P. O. Box 43325, La Tijera Station, Los Angeles, CA 90043

*Metri-Q:* Teaches children the metric units of length, prefixes of the metric units, and number relationships among the units. $4.00

## C. WHERE TO LOOK, WHAT TO READ

### References

American Association of School Librarians and National Council of Teachers of Mathematics. *One to Get Ready: A Selected Bibliography on Metrication.* Chicago: the Association, 1973. A brief listing of sources of metrication aids, suggested films and books for review, articles in professional journals, and selected government publications and materials for parents.

Brandou, Julian R. *A Collection of Materials and Ideas On . . . Metric Education.* East Lansing, MI: Michigan State University, 1974.

A collection of articles on metric education and teaching measurement; includes an extensive bibliography.

ERIC Information Analysis Center for Science, Mathematics and Environmental Education. *Materials for Metric Instruction.* Columbus, OH: Ohio State University, 1975.
A bibliography of approximately 125 items of supplementary materials: metric kits, task cards, films, filmstrips, slides, and miscellaneous items.

Hopkins, Robert A. *The International Metric System and How It Works.* Tarzana, CA: Polymetric, 1973.
A history of the metric system and its present status. Benefits and costs of the system, NBS information on SI units, and numerous tables and conversion factors are given.

*Metric Editorial Guide.* Second Edition. Washington, DC: American National Metric Council, 1975.
A guide to the proper writing and usage of metric terms with suggested American practice for using and punctuating them.

*Metric Manual.* Neenah, WI: J. J. Keller & Associates, 1974.
Information about conversion to the metric system of measurement for business, school, home, and library use. The history of measurement and metrication, pros and cons of metrication, standards and measurement comparisons, and numerous other items including a comprehensive glossary, references, and metric illustrations and charts are covered.

*Metric System Guide Library.* Five volumes. Neenah, WI: J. J. Keller, 1974.
I. Metrication in the United States: Orientation and Structure; II. Legislation and Regulatory Controls: Federal and All States; III. Metric Units, Comparisons, Factors, Tables; IV. Metric Definitions; V. Reference Sources.

*Metric Text.* New York: Central Instrument Co., December 1973.
A fully illustrated text dealing with the history of metrics, understanding and using the metric system in everyday life, metrics in drafting, conversion factors, conversion tables, and a guide to available aids.

Murphy, M. O. and Polzin M. A. "A Review of Research Studies on the Teaching of the Metric System." *Journal of Educational Research,* February 1969, pp. 267–270.

Research studies on the teaching of the metric system and measurement are reviewed.

National Council of Teachers of Mathematics. *NCTM Metrication Update and Guide to Suppliers of Metric Materials.* Washington, DC: The Council, 1975.

A listing of about 200 suppliers of manipulative aids, A-V aids, books, workbooks, charts and posters, kits, reports and pamphlets, periodicals, and games.

Page, Chester H. and Vigourerex, Paul (eds.) *The International System of Units (SI).* National Bureau of Standards, Special Publication 330, 1974.

The English version of the international resolutions from 1889 to 1971, including the agreements defining "Le Systeme International d'Unités" — the SI.

*Some References on Metric Information, With Charts On: All You Need To Know About Metric (And) Metric Conversion Factors.* National Bureau of Standards, Special Publication 389. Revised 1975.

A list of publishers and location of specific metric materials; includes books, kits, posters, other instructional materials.

U. S. Department of Commerce, National Bureau of Standards. *NBS Guidelines for Use of the Metric System.* Washington, DC: Bureau of Standards, 1974.

A list of publications of the Bureau of Standards and the American National Standards Institute, organizations that market metric materials for educators, and additional sources of information.

U. S. Department of Commerce, National Bureau of Standards. *U. S. Metric Study Interim Reports.* Washington, DC: Government Printing Office, 1971.

Findings of the U. S. Metric Study requested by Congress are detailed in 12 reports: *International Standards; Federal Government: Civilian Agencies; Commercial Weights and Measures; The Manufacturing Industry; Nonmanufacturing Industry; Education; The Consumer; International Trade; Department of Defense; A History of the Metric Controversy in the United States; Engineering Standards; Testimony of Nationally Representative Groups.* Also included is a summary report (see DeSimone, *A Metric America,* under General, Historical, and Professional Background for the Teacher).

**General, Historical and Professional Background for the Teacher**

Adams, Herbert F. R. *SI Metric Units: An Introduction.* New York: McGraw-Hill, 1974.

A short history, explanations of units, problems, solutions, and conversion tables are given.

Alexander, F. D. "The Metric System — Let's Emphasize Its Use In Mathematics." *Arithmetic Teacher.* May 1973, pp. 395–396.

A brief look at the development of the English and metric systems of measurement. The advantages and disadvantages of the metric system are discussed.

Ambler, Ernest. "Measurement Standards, Physical Constants, And Science Teaching." *The Science Teacher.* November 1971, pp. 63–71.

A description of the development of the metric system and how physical measurements and standards are built up from a few base units. Most of the paper deals with physics topics.

*An Introduction to the SI Metric System.* Sacramento, CA: California Department of Education, 1975.

An inservice guide for teaching measurement, kindergarten through grade eight. Curriculum strategies and classroom activities are discussed. The history and salient features of SI metrics are presented.

Biggs, Edith. "Metrication in the School Curriculum." *Trends In Education.* April 1972, pp. 35–40.

Problems initially faced in the British metrication process are discussed.

Bowles, D. Richard. "The Metric System in Grade 6." *Arithmetic Teacher.* January 1964, pp. 36–38.

Fifteen concepts and generalizations in the unit, "The Metric System of Measurement," developed by the Austin Public Schools for sixth grade classes, are listed.

Bright, George W. "Bilingualism in Measurement: The Coming of the Metric System." *Arithmetic Teacher.* May 1973, pp. 397–399.

The responsibility of teachers of developing in their students a bilingualism in the English and metric languages of measurement is discussed.

Chalupsky, Albert B., Crawford, Jack J., and Carr, Edwin M. *Going Metric: An Analysis of Experiences in Five Nations and Their Implications for U. S. Educational Planning.* Palo Alto, CA:

American Institutes for Research in the Behavioral Sciences, 1974.

Specific problems in metric education are identified and coping strategies are described. Nine recommendations for the United States are made.

Chalupsky, A. B. and Crawford, J. J. "Preparing the Educator to Go Metric." *Phi Delta Kappa.* December 1975, pp. 262–265.

Recent history of "going metric," policy issues relating to "going metric," and problems in teaching SI are discussed. Future metric projects are recommended.

Copeland, R. W. *How Children Learn Mathematics.* New York: Macmillan Co., 1970.

The standard topics on methods of teaching mathematics in elementary education are covered. The material is based on the work of Jean Piaget.

Deming, Richard. *Metric Power: Why and How We Are Going Metric.* New York: Thomas Nelson, 1974.

A general discussion of the issues involved in metrication.

DeSimone, D. V. *A Metric America: A Decision Whose Time Has Come.* National Bureau of Standards, 1971.

Illustrated report to the Congress from the Department of Commerce; history, practices, trends, recommendations, and information on metrication in other countries are given.

Drake, Paul. "Hello Metrics!" *Teacher.* October 1974, pp. 46–50.

The author tries to answer ten questions often asked about teaching metrics.

Dubisch, Roy. "Some Comments on Teaching the Metric System." *Arithmetic Teacher.* February 1976, pp. 106–107.

Suggestions are provided to save energy and money during the metric conversion period.

Edson, Lee. "New Dimensions for Practically Everything: Metrication." *American Education.* April 1972, pp. 10–14.

A history of metrics from the French Revolution to the present, with emphasis on Britain's metrication and its implications for the American changeover.

Firl, Donald H. "The Move to Metric: Some Considerations." *Mathematics Teacher.* November 1974, pp. 581–584.

A brief historical review of the metric system is given. The various SI units are discussed. Some of the problems involved in changing to a different system of measurement are given.

Gilbert, Thomas F. and Gilbert, Marilyn B. *Thinking Metric.* New York: John Wiley & Sons, Inc., 1973.
A programmed self-teaching guide.
Grzesiak, Katherine A. "America and the Metric System: Present Perspectives." *Elementary School Journal.* January 1976, pp. 195–199.
A discussion of the reasons for the changeover to metrics, recent history of metrics, and the effects of the changeover on pupils in the elementary schools.
Hallerberg, Arthur E. "The Metric System: Past, Present — Future?" *Arithmetic Teacher.* April 1973, pp. 247–255.
A brief history of measurement, leading to the decision to change to the metric system in the United States. The article is followed by a list of eight advantages of the metric system.
Hawkins, Vincent J. "Teaching the Metric System as Part of Compulsory Conversion in the United States." *Arithmetic Teacher.* May 1973, pp. 390–394.
Metric topics are suggested for each grade level from K to 6, the junior high school, and senior high school.
Helgren, Fred J. "The Metric System in the Elementary Grades." *Arithmetic Teacher.* May 1968, pp. 349–353.
A timetable for grades three to eight is suggested for teaching metrics as a system in itself.
Helgren, Fred J. "Schools Are Going Metric." *Arithmetic Teacher.* April 1973, pp. 265–267.
A six-point plan is discussed to introduce metrics into the schools, and the use of metrics in various careers is mentioned.
Higgins, Jon L (ed.) *A Metric Handbook for Teachers.* Reston, VA: The National Council of Teachers of Mathematics, 1974.
A collection of articles giving practical suggestions to teachers.
*Interstate Consortium on Metric Education: Final Report.* Sacramento, CA: California State Department of Education, 1975.
Twenty-three recommendations relating to instructional materials and pedagogy, implementation of changeover to metrics, and teacher education are made by a 28-member interstate consortium.
Izzi, John. *Metrication, American Style.* Bloomington, IN: Phi Delta Kappa Educational Foundation, 1974.
A short history and overview of the metric system, recommended sources and appropriate cautions are given.

Jones, Philip G. "Metrics: Your Schools Will Be Teaching It and You'll Be Living It — Very, Very Soon." *American School Board Journal.* July 1973, pp. 21–25.

A brief background of the metric system in the United States is given, followed by a school metrication calendar recommended for the 10-year conversion period. Expected changes in schools and school districts are included.

King, Irv, and Whitman, Nancy. "Going Metric in Hawaii." *Arithmetic Teacher.* April 1973, pp. 258–260.

The statewide program for converting Hawaii schools to the metric system is described.

Lindstedt, S. A. "Metrication." *Education Canada.* Spring 1976, pp. 16–19.

Issues pertinent to the elementary school curricula and metrics are addressed: basic methodology, structure versus utility, decimals versus common fractions, and precision and ragged decimals.

Miller, Byron S. and Trimbo, Henry B. "Use of Metric Measurements in Food Preparation." *Journal of Home Economics.* February 1972, pp. 20–25.

The advantages of changing from volume to weight measurements are discussed.

Moss, Jeanette K. "Teaching Aids: Tooling Up for the Metric Changeover." *Teacher.* March 1974, pp. 90–97.

Recently produced materials that may help teachers and students make the change smoothly from the English system of measurement to the metric system are listed.

Muellen, T. K. "Metric in Maryland." *Educational Leadership.* February 1974, pp. 435–437.

The Maryland State Board of Education plan and procedures for complete metrication are discussed.

National Council of Teachers of Mathematics. *Measurement In School Mathematics.* 1976 Yearbook. Washington, DC: The Council, 1976.

Essays on current issues and interests covering the whole spectrum of measurement and its role in school mathematics: learning and teaching measurement, selected measurement activities, and selected measurement resources for teaching.

National Council of Teachers of Mathematics, Metric Implementation Committee. "Metric Competency Goals." *Mathematics Teacher.* January 1976, pp. 90–91.

A list of 28 metric competencies are suggested as a basic guide for planning instructional programs. The competencies are categorized as those expected at the end of the third, sixth, and ninth grades.

National Council of Teachers of Mathematics, Metric Implementation Committee. "Metrics: Not If, But How." *Arithmetic Teacher.* May 1974, pp. 366–369.
Some general guidelines for teaching measurement and specific guidelines for teaching the metric system are discussed.
Nation's Schools. "Students Learn to Live with Liters and Meters." *Nation's Schools.* April 1974, pp. 24–25.
How particular teachers developed and implemented a metric program for elementary grades is discussed.
Piaget, Jean. "How Children Form Mathematical Concepts." *Scientific American.* November 1953, pp. 74–79.
How children form mathematical concepts is shown through a series of experiments.
Ritchie-Calder, Lord "Conversion to the Metric System." *Scientific American.* July 1970, pp. 17–25.
Traces the development of the empirical system of measurement, and surveys the history of the metric system and the International System of Units (SI). The decision to convert to metrics in Britain and British metrication are discussed.
Robinson, Berol D. *Education (An Interim Report of the U. S. Metric Study).* National Bureau of Standard Special Publication 345-7. 1971.
A status report to the Congress from the Department of Commerce that also contains recommendations for education.
Schreiber, Edwin W. "Significant Facts in the History of the Metric System for Teachers of Junior and Senior High School Mathematics." *Mathematics Teacher.* November 1969, pp. 373–381.
The general background of the metric system is described, followed by the history of metrics in the United States.
Sengstock, Wayne L. and Wyatt, Kenneth E. "Meters, Liters, and Grams." *Teaching Exceptional Children.* Winter 1976, pp. 58–65.
The metric system and its implications for curriculum for exceptional children are discussed.
Swan, Malcolm D. "Experience, Key to Metric Unit Conversion." *The Science Teacher.* November 1970, pp. 69–70.
The need for schools to provide students with direct experiences

in metric units rather than converting from English to metric units are discussed.

Treat, Charles F. *A History of the Metric System Controversy in the United States.* National Bureau of Standards, Special Publication 345-10. 1971.

An account of the metric system controversy based on extensive survey of historical data.

Vervoot, Gerardus. "Inching Our Way Towards the Metric System." *Arithmetic Teacher.* April 1973, pp. 275–279.

General information on the development of measurement is followed by a brief history and explanation of the units of measure in the metric system. The United States' efforts to go metric are also discussed.

Warning, Margaret. "Start Now to Think Metric." *Journal of Home Economics.* December 1972, pp. 18–21.

Stages through which people advance before using a new product (or system) with the east of habit, the experiences of other countries in the metric changeover, and the metric system in the United States are discussed.

### Ideas and Activities for the Elementary Classroom

*An Educator's Guide to Teaching Metrication.* Chicago: Sears, Roebuck, revised edition, 1975.

Free upon request to Sears Consumer Information Services. History, objectives, and interdisciplinary activities and projects are included.

Armbruster, F. O. *Think Metrics: A Basic Guide to the Metric System.* San Francisco: Troubador Press, 1974.

Excellent graphics relating everyday items to metric measures.

Ashlock, Robert B. "Introducing Decimal Fractions with the Meterstick." *Arithmetic Teacher.* March 1976, pp. 201–206.

Seven activities that provide students with experiences that help them understand the decimal representation of fractions.

Barnett, Carne. *Metric Ease.* Palo Alto, CA: Creative Publications, 1975.

A collection of over 70 activities.

Bolster, Carey L. "Activities: Centimeter and Millimeter Measurements." *Mathematics Teacher.* November 1974, pp. 623–626.

Reproducible activity sheets for classroom use in estimating and measuring to the nearest centimeter and millimeter are provided. Students' worksheets provide a metric ruler and a metric caliper to be assembled.

Bright, George, W. "Metrics, Students and You!" *Instructor.* October 1973, pp. 59–66.
Activities to introduce the most common metric units into the elementary curriculum are made.

Bruni, James V. and Silverman, Helene. "An Introduction to Weight Measurement." *Arithmetic Teacher.* January 1976, pp. 4–10.
Numerous readiness activities for mass measurement.

Bruni, James V. and Silverman, Helene. "Organizing a Metric Center in Your Classroom." *Arithmetic Teacher.* February 1976, pp. 80–87.
A description of a teacher-made metric center including things to measure, standard measuring instruments, activity cards, and other supplies needed for the measurement activities. Activities for linear, area, volume or capacity, and mass measurement are included.

Buckeye, Donald A. *I'm OK, You're OK, Let's Go Metric.* Troy, Michigan: Midwest Publications, 1973.
A collection of classroom activities that are divided into four reading levels: no read, low read, medium read, and high read. The pages are in "tear-out" form.

Burton, Grace. "A Potpourri of Metric Activities." *Elementary School Journal.* January 1976, pp. 201–207.
A group of practical (almost no cost and little preparation time) metric activities for the elementary school classroom.

Cech, Joseph P. and Seltzer, Carl. *Working with Color Rods in Metric Measurement; Metric Length, Metric Area, and Metric Volume.* Skokie, IL: National Textbook Company, 1973.
Three booklets of duplicating masters on measuring length, area, and volume.

Clack, Alice A. and Leitch, Carol. *Amusements in Developing Metric Skills.* Troy, MI: Midwest Publications, 1973.
A collection of various puzzles such as dots-to-dots, line designs, and code puzzles. The pages are in "tear-out" form.

Crane, Beverly. "The Metrics Are Coming." *Grade Teacher.* February 1971, pp. 88–89 and 154.

Describes introducing the terminology and mathematical organization of the metric system to third graders in a way that they can enjoy and understand.

Doherty, Joan. "Getting a Good Start in Teaching Metric Measurement Meaningfully." *Arithmetic Teacher.* May 1976, pp. 374–378.

Activities for developing metric concepts are described. Included are "starting point" activities, games, bulletin board displays, and action projects to educate others.

Fisher, Ron. "Metrics Is Here; So Let's Get On With It." *Arithmetic Teacher.* May 1973, pp. 400–402.

Classroom activities involving investigation, observation, and measurement skills are suggested.

Freeman, William W. K. "Think Metric About Weather." *Arithmetic Teacher.* May 1975, pp. 378–381.

Teaching metrics using readings on a Celsius scale and seasonal temperature changes is suggested.

Geer, Charles P. and Geer, John W. *MathMETRICS.* Burlingame, CA: Modern Math Materials, 1975.

A book of activities for teaching the metric system for grades three to eight. Most of the activities are the paper and pencil type.

Great Ideas. *Metric Arithmapuzzles.* Commack, NY: Great Ideas, 1973.

A set of crossnumber puzzles dealing exclusively with the metric system.

Gould, Carole. *Color Me Metrics.* San Jose, CA: A. R. Davis and Co., 1973.

A reproducible coloring and measuring book covering linear measurements. Suitable for primary grades.

Hallamore, Elizabeth. *The Metric Book of Amusing Things to Do.* Woodbury, NY: Barron's Educational Series, Inc., 1974.

A collection of games, puzzles, and projects designed to give children practical experience with the metric system.

Hampel, Paul and Stein, Bill. "Step Right Up to the Merry Metric Carnival." *Teacher.* March 1976, pp. 52–54.

An all-day, all-school carnival designed to help students gain insight into the practicality and versatility of the metric system is described.

Henderson, George L. and Glunn, Lowell D. *Let's Play Games in Metrics.* Skokie, IL: National Textbook, 1974.
A collection of games and activities for teaching the metric system.

Henry, Boyd. *Teaching the Metric System.* Chicago: Weber Costello, 1973.
Hands-on activities and materials are suggested. Easily duplicated charts, puzzles, and games are included.

Holmberg, Verda. *The Metric System of Measurement.* Los Gatos, CA: Contemporary Ideas.
A workbook of enrichment experiences for upper elementary students with linear, area and volume, liquid capacity and weight, and temperature measurements.

Immerzeel, George, and Wiederanders, Don. "Ideas." *Arithmetic Teacher.* April 1973, pp. 280–287.
Activities for levels one to eight are given. The activities use the number line to relate measures within the metric system and to relate metric and English measures. Reproducible activity sheets are included.

Lampman, Donna. "Metrication of the American Family." *Arithmetic Teacher.* Dec. 1974, pp. 707–709.
A delightful and informative metric awareness assembly produced by fifth graders is described. The assembly contains many ideas for the classroom teacher.

Leffin, Walter. *Going Metric: Guidelines for the Mathematics Teacher, Grades K–8.* Washington, DC: NCTM, 1975.
A brief history of the metric system, a detailed explanation of the major SI units, and general guidelines for teaching SI are given. Detailed classroom activities, lists of recommended materials, and instructions for student-made learning aids are included.

Long, Betty and Witte, Carol. *Fun with Metric Measurement.* Manhattan Beach, CA: Teachers, 1973.
A collection of activities for grades three to nine with emphasis on doing. The activities cover length, volume, mass, area, and temperature.

Long, Betty and Witte, Carol. *Early Childhood Metric Fun Book.* Manhattan Beach, CA: Teachers, 1974.
A collection of activities for kindergarten to grade three children with emphasis on doing. Copies may be reproduced for classroom use.

The Math Group. *Metric Measurement: Activities in Capacity, Mass, and Temperature.* Minneapolis: The Math Group, 1974.
Answers to exercises are used in puzzles to provide the motivation for practice and drill.

The Math Group. *Metric Measurement: Activities in Linear Measurement.* Minneapolis: The Math Group, 1974.
Answers to exercises are used in puzzles to provide motivation for practice and drill.

Milber, Mary and Richardson, Toni. *Merry Metric Cookbook.* Hayward, CA: Activity Resources, 1974.
A collection of cooking-in-metric recipes for elementary pupils.

Miller, Mary Richardson. *Making Metric Maneuvers.* Hayward, CA: Activity Resources Company, 1974.
A collection of aids for teachers and classroom-tested activities for children involving physical movements.

Morehouse, T. and Schoonmaker, E. "Metric Month at Taft Middle School." *Phi Delta Kappan.* December 1975, p. 265.
A description of how a middle school gave its students a working knowledge of the metric system that was interdisciplinary in nature.

National Science Teachers Association. *Metric Exercises: Lively Activities on Length, Weight, Volume, and Temperature.* Washington, DC: the Association, 1973.
A booklet of exercises and activities for learning the fundamentals of the metric system designed for elementary through senior high school students.

Odegard, Sharon E. *World of Metric.* Canada: Ginn and Company, 1975.
Activities workbook to help students understand what measurement is, why the metric system was chosen, and how it works, and to provide practice to help students "think metric." Throughout the workbook emphasis is on estimating before measuring.

Phenomena for Inquiry Series. "Think Metric." *Science and Children.* March 1975, p. 8.
Questions that deal with everyday things and events relative to metrics to initiate creative research by children are given.

Peavler, Cathy Seeley. "Metricating Painlessly, Cheaply, Cooperatively." *Arithmetic Teacher.* October 1974, pp. 533–536.
Suggestions are provided for use in the classroom, for finding ma-

terials and for involving community in the teacher's metrication program. Projects for metrication and metric measuring aids that the teacher can make at no cost are suggested.

Pottinger, Barbara. "Measuring, Discovering, and Estimating the Metric Way." *Arithmetic Teacher.* May 1975, pp. 272–277. A teacher of grade three shares her ideas on teaching a metric unit. Noteworthy are the games she or her students have developed.

Ring, Art. *Metric Education Handbook.* Englewood Cliffs: Prentice-Hall Learning Systems, 1974. A practical, "how to," and technique book. It covers teaching strategies, teaching tools, learning stations, and conversion information.

Rucker, Isabelle P. "The Metric System in Junior High School." *Mathematics Teacher.* December 1958, pp. 621–623. An eighth grade math unit on the metric system of measurement is presented. The necessary materials are listed, and some of the classroom activities used are described.

Sheffield, Susan S. "The Meter Stick." *Science and Children.* March 1973, pp. 22–24. A detailed discussion of how a teacher made it possible for her fourth graders to experience the metric system.

Shumway, Richard J., and Sachs, Larry. "Don't Just Think Metric — Live Metric." *Arithmetic Teacher.* February 1975, pp. 103–110. Various metric activities used at teacher workshops are described. Materials used are inexpensive and readily available.

Silvani, Harold. *Riddles: Number Puzzles Book A.* Fresno, CA: Creative Teaching Associates, 1974. A collection of 24 worksheets involving computation with metric measures. Students are asked to complete the problems, use code letters, and discover the answer to a riddle.

Strangmen, K. B. "Grids, Tiles, and Area." *Arithmetic Teacher.* December 1968, pp. 668–672. A mathematically sound and teachable method for introducing area to fifth grade pupils is described.

Trueblood, Cecil R. *Metric Measurement: Activities and Bulletin Boards.* Dansville, NY: Instructor Curriculum Materials, 1974. General guidelines for teaching measurement; an explanation of metric; and learning activities for linear, area, weight, and volume are provided.

Trueblood, Cecil R., and Szabo, Michael. "Procedures for Designing Your Own Metric Games for Pupil Involvement." *Arithmetic Teacher.* May 1974, pp. 404–408.
Seven criteria for designing metric games are presented and discussed.
West, Tommie A. "Teaching Metrics to Beginners." *Today's Education.* November–December 1974, pp. 80–82.
Ideas that primary teachers can use in teaching are described.
Viets, Lottie. "Experiences for Metric Missionaries." *Arithmetic Teacher.* April 1973, pp. 269–273.
Activities and materials for measuring and estimating are suggested, and charts and conversion graphs that can be used are shown.
Wray, D. Eileen. "You and the Metric System." *Arithmetic Teacher.* December 1964, pp. 576–580.
The metric units of length, weight, and volume are introduced in story form.

**Children's Books**

Bates, W. W. and Fullerton, O. *How to Think Metric.* Toronto, Canada: Copp Clark Publishing, 1974.
The metric system is explained in simple terms. The authors explain how to use the units you will need.
Behrens, June. *True Book of Metric Measurement.* Chicago, IL: Children's Press, 1975.
A brief introduction to the metric system for grades one to four.
Bendick, Jeanne. *How Much and How Many; The Story of Weights and Measures.* New York: McGraw-Hill, Inc., 1960.
The ways of measuring and counting in commerce and science are discussed and illustrated.
Branley, Franklyn M. *Measure With Metric.* New York: Thomas Y. Crowell, 1975.
An activity approach to metric measures of length, mass, and capacity. Colorful illustrations. Simple, familiar, everyday things are measured.
Branley, Franklyn. *Think Metric.* New York: Crowell, 1972.
A colorful book in narrative style, with historical and descriptive data about metric and English customary system of measurement.

Engel, William C. "Are You Ready to Go Metric?" *Mathematics Student.* October 1973, number 1, pp. 1–3.
A brief introduction to the metric system, including lengths, capacity, weight, and temperature. A quiz is given to help readers determine if they are ready for a metric society.

Friskey, Margaret. *About Measurement.* Chicago, Melmount Publishers, Inc., 1965.
Colorful illustrations and discussion about the historical units of measurement.

Hahn, James and Hahn, Lynn. *The Metric System.* New York: Franklin Watts, Inc., 1975.
A history of the development of the metric system for grades three through seven.

Hiatt, Mary and Harvey, Lina. *The Metric Mice Measure Weight, The Metric Mice Measure Length, The Metric Mice Measure Area, The Metric Mice Measure Volume.* Inglewood, CA: Educational Insights, 1974.
Devoted to general and metric measurements.

Hirsch, S. Carl. *Meter Means Measure: The Story of the Metric System.* New York: Viking, 1973.
Reasons for going metric are discussed through the historical development of measurement and the metric system.

McDonald's Corporation. *Fun Course in McMetrics.* Chicago: The Corporation, 1974.
A free booklet presenting a metric unit on length. Puzzle activities are included.

Nation, Kay and Nation, Bob. *Meters, Liters, and Grams.* New York: Hawthorn Books, Inc., 1975.
The metric system presented as a reader. Appropriate for the intermediate grades.

Ross, Frank, Jr. *The Metric System: Measures for All Mankind.* ill. by Robert Galster. New York: S. G. Phillips, 1974.
The origin and development of the metric system to the present SI is traced. Tables, conversion charts, and glossary are included.

Willert, Fritz. *My Metric Measurement Manual.* Two Rivers, WI: Pauper Press, 1974.
A cartoon story introduces the metric system to children.

Zim, H. S. and Skelly, J. R. *Metric Measure.* New York: Morrow, 1974.

The importance of standardized measurement is shown in a well-illustrated book. Familiar and interesting aspects of everyday life are expressed in metric terms. The basic units of metric measurement are discussed.

## For Parents

Ascher, William. "Just a Silly Little Millimeter." *Plan and Print.* April 1973, pp. 12–13, 23.
   The effects of metrication in the United States on industry, engineering, and the standards used are discussed.
Barbow, Louise E. *What About Metric?* U. S. Department of Commerce, National Bureau of Standards, Consumers Information Series 7. Washington, DC: Government Printing Office, 1973.
   Consumer information about the metric system in the marketplace, in everyday use, and for workers.
Batcher, Olive M., and Louise A. Young. "Metrication and the Home Economist." *Journal of Home Economics.* February 1974, pp. 28–31.
   The fields of housing and equipment, clothing and textiles, and food and food preparation are affected by the metric conversion. Metric activities for the classroom are listed. A recipe for vanilla cake using metric measures is also given.
Channing L. Bete Co., Inc. *Going Metric . . . It's The Only Way To Go.* Greenfield, MA: Channing L. Bete Co., Inc., 1974.
   A booklet of sketches on the metric system and why America is going metric.
Cortright, Richard W. "Adult Education and the Metric System." *Adult Education.* November 1971, p. 190.
   The need for adult education to be included in the long-range plan for the conversion to metrication is discussed.
Donovan, F. *Let's Go Metric.* New York: Weybright and Talley, 1974.
   The metric system is explained to students and the general American public; and the effects the changeover will likely have in our everyday life are discussed.
Glaser, Anton. *Neater by the Meter: An American Guide to the Metric System.* Southampton, PA: the Author (1237 Whitney Rd., 19866), 1974.

Written for the nontechnical person and oriented to everyday needs.

Kendig, Frank. "Coming of the Metric System." *Saturday Review.* November 25, 1972. pp. 40–44.

Provides general information on the English and metric systems. Conversion formulas and tables of equivalents are given. Illustrations with explanations on each of the basic metric units for length, mass, volume, and capacity are included.

Holt, Susan F. *The United States and the Metric System.* Exponent Series. Minneapolis: Federal Reserve Bank of Minneapolis, 1973. Looks at worldwide standardization and its implications for world trade. Reviews the history of metrication, pros and cons of its decimal base and the task of conversion. Contains bibliography that reflects the interest of the business world.

*Moving Toward Metric.* New York: J. C. Penney, 1974.

Free on request to Educational Relations Department. A packet of material on metric consumerism. Scripts for radio-TV are included.

Rothrock, B. D. *The Consumer (An Interim Report of the U. S. Metric Study.)* National Bureau of Standards, Special Publication 345-7. 1971.

One of 12 reports to the Congress by the Department of Commerce. This one is based on a study of consumer attitudes and issues.

Schimizzi, Ned V. *Mastering the Metric System.* New York: New American Library, 1975.

A brief history of the metric system; definitions, equivalents, and notation of every unit in SI; extensive table of equivalents and conversion factors.

Stover, Allan C. *You and the Metric System.* New York: Dodd, Mead & Co., 1974.

Discusses what the metric units are, how they are used, how the changeover will affect the public, and the problems involved.

*The Metric Song.* Filmstrip and cassette, color. J. C. Penney Company, 1974.

Available on loan through J. C. Penney stores, Educational Relations Department. Entertaining explanation of why we are going metric. Musical listings of metric units, uses, and relationships.

*The Swing to Metric.* Detroit: General Motors Corp., 1973.

Request pamphlet from Personnel Communications Department. Background and growth of the metric system with applications to General Motors is covered.

Willens, Michele. "U. S. Moving, Inch By 2.54 cm, To Metric System." *Mainliner* (United Air Lines), July 1973, pp. 24–31. The ease and efficiency of the metric system is noted. Companies such as IBM and GM are incorporating metrics into their systems, and Ohio is using metric as well as standard road signs in their efforts to change. The implications that changing to metrics has for the housewife, businessman, schools, and labor unions are also discussed.

## For Additional Information

### Organizations

American National Metric
  Council
1625 Massachusetts Ave., N. W.
Washington, DC 20036

Center for Metric Education
Western Michigan University
Kalamazoo, MI 49001

National Bureau of Standards
U. S. Department of Commerce
Washington, DC 20234

National Council of Teachers of
  Mathematics
1906 Association Dr.
Reston, VA 22091

National Science Teachers Association
1201 16th St., N. W.
Washington, DC 20036

U. S. Metric Association
Sugarloaf Star Route
Boulder, CO 80302

### Newsletters and Journals

*The Arithmetic Teacher,* National Council of Teachers of Mathematics, 1906 Association Dr., Reston, VA 22091

*The Mathematics Teacher,* National Council of Teachers of Mathematics, 1906 Association Dr., Reston, VA 22091

*The Science Teacher,* National Science Teachers Association, 1742 Connecticut Ave., N. W., Washington, DC 20009

*American Metric Journal,* AMJ Publishing Company, Drawer L, Tarzana, CA 91356

*Metric Association Newsletter,* Sugarloaf Star Route, Boulder, CO 80302

*Metric News,* P. O. Box 248, Roscoe, IL 61073

*Metric Reporter,* American National Metric Council, 1625 Massachusetts Ave., N. W., Washington, DC 20036

*Metric System Guide Bulletin,* J. J. Keller and Associates, Inc.,
Neenah, WS 54956

*AIR Metric-gram,* AIR Metric Study Center, P. O. Box 1113, Palo
Alto, CA 94302

# appendix
## iii

# SAMPLE
# CLASSROOM
# ACTIVITIES

## LENGTH

### Paper Magic

You need:  adding machine tape
scissors
tape or glue
centimeter tape or ruler

1. Cut off two 50 cm long pieces of adding machine tape.
2. Fold each piece lengthwise.
3. Unfold and cut on the fold line of each paper tape. You now should have four pieces of paper tape.
4. a. Take one of the paper tapes and glue or tape the ends together. Overlap the ends about 1 cm.
   b. Take a second tape. Give it a half twist before taping the ends together. (See Figure III. 1.)
   c. Take a third paper tape. Before taping the ends together give it two half-twists.
   d. Take your last paper tape. Give it three half-twists before taping its ends.
5. a. Cut your first paper tape down the middle of the paper. (See Figure III. 2.)
   b. Cut your second and third paper tapes in the same way. What happened?
   c. Guess what will happen when you cut your last paper tape in the same way. Cut it and see!

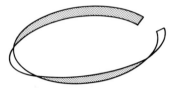

Figure III. 1  A half twist.

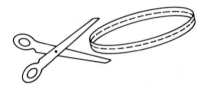

Figure III. 2  Cut down the middle.

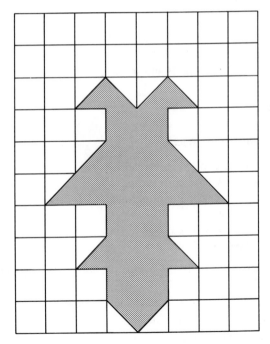

Figure III. 3  A possible shape.

## AREA

### Take Note!

You will need:  construction paper
square centimeter grid paper
scissors
pen or pencil
miscellaneous items to decorate notecards (optional)

1.  Use your square centimeter grid paper to make notecard shapes with the following areas:
Shape A:  100 square centimeters
Shape B:  20 square centimeters
Shape C:  40 square centimeters
(See Figure III. 3.)

2.  Cut out and place shape A over the construction paper.  Mark the outlines.

3.  Cut the construction paper on the outlines.

4.  Design or write a message on your cutout construction paper.

5.  Repeat, using shape B and shape C.

## CAPACITY

### Hawaiian Mint Delight Recipe

Ingredients:  60 mℓ softened margarine
80 mℓ light corn syrup
5 mℓ peppermint extract
2 mℓ salt
1 ℓ powdered sugar
green food coloring

1.  In a large bowl mix the margarine, corn syrup, peppermint extract, and salt.

2.  Gradually add the powdered sugar to the mixture.

3.  Divide the mixture in half.
a. Leave one half alone.
b. Add a drop of green food coloring to the other half.  Knead coloring into the mixture.

4.  Shape mixture into round balls about 2 cm across.  Place on wax paper.  Flatten balls with the palm of your hand.

5.  Let dry for a few hours.

6.  Makes about 35 candies.


## MASS

### Making Play-Clay

You will need:   170 g table salt
                         70 g flour
                         about 50 mℓ water
                         food color (optional)

Procedure:
1.  Put the salt and flour in a big bowl.

2.  Add food color to water (optional) and gradually add the water to the flour and salt.

3.  Add more water if necessary to make the clay workable.

4.  Make many cubic centimeters.

5.  Use your cubic centimeters to make objects of your choice. Keep track of the number of cubic centimeters you use in making the objects.

6.  Let objects harden.


## TEMPERATURE AND CAPACITY

### Kanten (Japanese Gelatin)

Ingredients:   350 mℓ tap water (about 25°C)
                       4 envelopes of plain gelatin
                       2 small packages of jello, any flavor
                       800 mℓ very hot water (about 100°C)
                       about 20 mℓ of lemon juice
Yield:   about 2.5 cm X 20 cm X 20 cm pan

1.  In a bowl mix gelatin and water until dissolved.

2.  In another bowl mix together jello, sugar, hot water, and lemon juice.

3. Pour contents of both bowls into pan.  Stir.
4. Refrigerate (about $10^\circ$C).
5. Cut into cubes and share with the entire class.

# INDEX